Introduction to

REAL-TIME IMAGING

SPIE PRESS Tutorial Texts Series

Covering fundamental topics in optical science and engineering at the introductory and intermediate levels.

An Introduction to Nonlinear Image Processing, *by Edward R. Dougherty and Jaakko Astola*
Vol. TT16, ISBN 0-8194-1560-X

Integration of Lasers and Fiber Optics into Robotic Systems, *by Janusz A. Marszalec and Elzbieta A. Marszalec*
Vol. TT17, ISBN 0-8194-1313-5

Sensor and Data Fusion Concepts and Applications, *by Lawrence A. Klein*
Vol. TT14, ISBN 0-8194-1227-9

Image Formation in Low-Voltage Scanning Electron Microscopy, *by L. Reimer*
Vol. TT12, ISBN 0-8194-1206-6

An Introduction to Morphological Image Processing, *by Edward R. Dougherty*
Vol. TT9, ISBN 0-8194-0845-X

An Introduction to Optics in Computers, *by Henri H. Arsenault and Yunlong Sheng*
Vol. TT8, ISBN 0-8194-0825-5

Digital Image Compression Techniques, *by Majid Rabbani and Paul W. Jones*
Vol. TT8, ISBN 0-8194-0648-1

An Introduction to Biological and Artificial Neural Networks for Pattern Recognition, *by Steven K. Rogers and Matthew Kabrisky*
Vol. TT4, ISBN 0-8194-0534-5

Aberration Theory Made Simple, *by Virendra N. Mahajan*
Vol. TT6, ISBN 0-8194-0536-1

Single Frequency Semiconductor Lasers, *by Jens Buus*
Vol. TT5, ISBN 0-8194-0535-3

IEEE PRESS Understanding Science & Technology Series

The IEEE PRESS Understanding Science and Technology Series treats important topics in science & technology in a simple and easy-to-understand manner. Designed expressly for the nonspecialist engineer, scientist, or technician, as well as the technologically curious—each volume stresses practical information over mathematical theorems and complicated derivations.

Understanding the Nervous System, *by Sid Deutsch and Alice Deutsch*
IEEE Order No. PP2915

Understanding Digital TV, *by Brian Evans*
IEEE Order No. PP4366

Understanding Telecommunications and Lightwave Systems, Second Edition, *by John G. Nellist*
IEEE Order No. PP4665

Understanding Lasers, Second Edition, *by Jeff Hecht, Sr.*
IEEE Order No. PP3541

Understanding Electro-Mechanical Engineering, *by Larry Kamm*
IEEE Order No. PP3806

Understanding Object-Oriented Engineering, *by Stefan Sigfried*
IEEE Order No. PP4507

Introduction to

REAL-TIME IMAGING

Edward R. Dougherty
Center for Imaging Science
Rochester Institute of Technology

Phillip A. Laplante
Fairleigh Dickinson University

TUTORIAL TEXTS IN OPTICAL ENGINEERING
Volume TT19
Donald C. O'Shea, Series Editor
Georgia Institute of Technology

SPIE OPTICAL ENGINEERING PRESS

A Publication of SPIE—The International Society for Optical Engineering
Bellingham, Washington USA

The Institute of Electrical and Electronics Engineers, Inc., New York

Library of Congress Cataloging-in-Publication Data

Dougherty, Edward R.
Introduction to real-time imaging / Edward R. Dougherty, Phillip A. Laplante.
p. cm. — (Tutorial texts in optical engineering ; v. TT 19)
ISBN 0-8194-1789-0 (softcover)
1. Image processing—Digital techniques. 2. Real-time data processing.
I. Laplante, Phil. II. Title. III. Series.
TA1637.D683 1995
621.36'7'0285433—dc20 94-44221
CIP

ISBN 0-8194-1789-0

Copublished by
SPIE—The International Society for Optical Engineering
P.O. Box 10
Bellingham, Washington 98227-0010
Phone: 360/676-3290 Fax: 360/647-1445

IEEE Press
445 Hoes Lane
Piscataway, NJ 08855-1331
Phone: 1-800-678-IEEE Fax: 908/981-8062
IEEE Order Number: PP5368

Printed in the United States of America

Introduction to the Series

The Tutorial Texts series was begun in response to requests for copies of SPIE short course notes by those who were not able to attend a course. By policy the notes are the property of the instructors and are not available for sale. Since short course notes are intended only to guide the discussion, supplement the presentation, and relieve the lecturer of generating complicated graphics on the spot, they cannot substitute for a text. As one who has evaluated many sets of course notes for possible use in this series, I have found that material unsupported by the lecture is not very useful. The notes provide more frustration than illumination.

What the Tutorial Texts series does is to fill in the gaps, establish the continuity, and clarify the arguments that can only be glimpsed in the notes. When topics are evaluated for this series, the paramount concern in determining whether to proceed with the project is whether it effectively addresses the basic concepts of the topic. Each manuscript is reviewed at the initial stage when the material is in the form of notes and then later at the final draft. Always, the text is evaluated to ensure that it presents sufficient theory to build a basic understanding and then uses this understanding to give the reader a practical working knowledge of the topic. References are included as an essential part of each text for the reader requiring more in-depth study.

One advantage of the Tutorial Texts series is our ability to cover new fields as they are developing. In fields such as sensor fusion, morphological image processing, and digital compression techniques, the textbooks on these topics were limited or unavailable. Since 1989 the Tutorial Texts have provided an introduction to those seeking to understand these and other equally exciting technologies. We have expanded the series beyond topics covered by the short course program to encompass contributions from experts in their field who can write with authority and clarity at an introductory level. The emphasis is always on the tutorial nature of the text. It is my hope that over the next five years there will be as many additional titles with the quality and breadth of the first five years.

Donald C. O'Shea
Georgia Institute of Technology

January 1995

To

John and Christopher

Contents

Preface

Real-time processing for digital imaging concerns efficient, deterministic implementation of algorithms whose inputs include digital images and whose outputs are digital images, numerical features, symbolic representations, or decisions. Practically, a real-time demand occurs when one is faced with composing an algorithm that must complete an imaging task within some given time frame. Computation bottlenecks appear in many forms and, to some extent, each requires its own real-time implementation via algorithm design, software, hardware, or a combination thereof; nevertheless, there are certain fundamental algorithms at the center of digital image processing and our focus is upon the structure, computation, and application of these algorithms. Specifically, we treat linear, matrix, and nonlinear algorithms that appear across a wide range of imaging applications. Among the applications discussed are noise suppression, edge detection, matched filtering, and data compression. Specific operations covered include linear convolution, the discrete cosine transform (DCT), the fast Fourier transform (FFT), the median filter, and the morphological gradient.

The imaging algorithms discussed tend to be computationally intensive, especially when directly implemented on standard sequential hardware. Various means of providing efficient computation are discussed. For instance, fast matrix transforms result from decomposing a matrix into a cascade of more easily computable matrices. Among the various hardware paradigms treated are pipelining, dataflow, and systolic arrays. Numerous aspects of programming languages are discussed, including parameter passing, recursion, typing, and exception handling. There is also a survey of commonly used languages and ways in which these contribute or do not contribute to real-time processing. Numerous optimization techniques such as loop unrolling and loop jamming are provided for avoiding unnecessary computation at run time.

Following an account of real-time issues in the first chapter, there is a chapter introducing the basic computer architectures that will play a role in finding hardware solutions for real-time imaging tasks. The chapter contains an introductory account of sequential processing in the von Neumann architecture. This has been included to provide those who are not familiar with assembly-level programming with the basic architectural concepts and terminology that will be used subsequently. The next three chapters treat linear, matrix transform, and nonlinear imaging algorithms. Actual applications are given and computational aspects of the algorithms are discussed. The final three chapters discuss three levels at which one can address real-time processing: parallel hardware, the programming language, and code optimization.

These solutions are applied to various computations in the previously discussed imaging algorithms.Our goal has been to write a text that can be used by those who are not necessarily experts in either computer science or digital image processing, but who need to become familiar with the kinds of computation bottlenecks common in digital imaging and with some of the ways in which efficient real-time processing can be affected. References have been provided for those who wish to pursue individual topics in more detail.

We acknowledge and offer our appreciation to all who assisted in preparation of the book. These include Y. Chen, C. Cuciurean-Zapan, J. Astola, and M. Rabbani, who contributed the figures and images; J. Handley, S. Wilson, and D. Sinha, who technically reviewed the manuscript; and E. Pepper, who provided editorial assistance and who guided the manuscript through production.

Introduction to

REAL-TIME IMAGING

Chapter 1

What is Real-Time Processing?

Consider a software system in which the inputs represent digital data from hardware such as imaging devices or other software system's and the outputs are digital data that control external hardware such as displays. The time between the presentation of a set of inputs and the appearance of all the associated outputs is called the **response time.** A **real-time system** is one that must satisfy explicit bounded response time constraints to avoid failure. Equivalently, a real-time system is one whose logical correctness is based both on the correctness of the outputs and their timeliness. Notice that response times of, for example, microseconds are not needed to characterize a real-time system – it simply must have response times that are constrained and thus predictable. In fact, the misconception that real-time systems must be "fast" is because in most instances, the deadlines are on the order of microseconds. But the timeliness constraints or deadlines are generally a reflection of the underlying physical process being controlled. For example, in image processing involving screen update for viewing continuous motion, the deadlines are on the order of 30 microseconds. In practical situations, the main difference between real-time and non-real-time systems is an emphasis on response time prediction and its reduction.

Upon reflection, one realizes that every system can be made to conform to the real-time definition simply be setting deadlines (arbitrary or otherwise). For example, a one-time image filtration algorithm for medical imaging, which might not be regarded as real-time, really is real-time if the procedure is related to an illness in which diagnosis and treatment have some realistic deadline. Because all systems can be made to look as if they were real-time, we refine the definition somewhat in terms of the system's tolerance to missed deadlines. For example, **hard real-time systems** are those where failure to meet even one deadline results in total system failure. In **firm real-time systems** some fixed small number of deadlines can be missed without total system failure. Finally, in **soft real-time systems** missing deadlines leads to performance degradation but not failure. Unless otherwise noted, when we say "real-time" throughout this tutorial, we mean hard real-time.

Another common misconception is that the study of real-time processing is really a non-issue because hardware is always getting faster. By throwing faster hardware at the problem deadlines can always be met. However, as we just stated, unless one can predict performance and hence bound response times, one can never be satisfied that deadlines are always being achieved. Moreover, faster hardware is not always available or suitable for certain applications.

Some feel that real-time performance is easy to achieve. As we hope to show in this tutorial, that is not always so, largely because most hardware and programming languages are not suitable for real-time demands.

1.1 Characteristics of Real-Time Systems

Real-time systems are often reactive and/or embedded systems. **Reactive systems** are those in which functionality is driven by ongoing, sporadic interaction with their environment, such as in virtual reality. **Embedded systems** generally do not have a generalized operating system interface and are used explicitly to control specialized hardware devices. For example, many imaging systems that reside in special hardware platforms, such as virtual reality, multimedia, and medical imaging, are embedded.

An important concept in real-time systems is the notion of an **event**, that is, any occurrence that results in a change in the sequential flow of program execution. Events can be divided into two categories: synchronous and asynchronous. **Synchronous events** are those that occur at predictable times such as execution of a conditional branch instruction or hardware trap. **Asynchronous events** occur at unpredictable points in the flow-of-control and are usually caused by external sources such as a clock signal. Both types of events can be signaled to the CPU by hardware signals.

There is an inherent delay between when an interrupt occurs and when the CPU begins reacting to it called the **interrupt latency**. Interrupt latency is due to both hardware and software factors. Interrupts may occur periodically (at fixed rates), aperiodically, or both. Tasks driven by interrupts that occur aperiodically are called sporadic tasks. Systems where interrupts occur only at fixed frequencies are called **fixed rate systems** and those with interrupts occurring sporadically are called **sporadic systems**. For example, many imaging systems involve updating a display at from 20 to 40 times per second. A fixed rate task of, say, 30 hertz might be assigned to perform the image update. On the other hand, a target acquisition algorithm may run only when a candidate target image is on hand.

Another characteristic of a robust real-time system is that it is deterministic. A system is said to be **deterministic** if for each possible state, and each set of inputs, a unique set of outputs and the next state of the system can be determined. In particular, a certain kind of determinism called **event determinism** means that the next state and outputs of the system are known for each set of inputs that trigger events. Thus, a system that is deterministic is event deterministic. While it would be difficult for a system to be deterministic only for those inputs that trigger events, this is plausible and so event determinism may not imply determinism. We are, however, only interested in pure deterministic systems. Finally, if in a deterministic system the response time for each set of outputs is known, then the system also exhibits **temporal determinism**. Each of these previous definitions of determinism implies that the system must have a finite number of states. This is a reasonable assumption to make in a digital computer system where all inputs are digitized to within a finite range. For any physical system there are certain states under which the system is considered to be "out of control" and the software controlling such a system must avoid these states. For example, in certain guidance systems for robots or aircraft, rapid rotation through a 180° pitch angle can cause a physical loss of gyro control. The software must be able to foresee and prepare for this situation or risk losing control. One side benefit of designing deterministic systems is that one can guarantee that the system can respond at any time, and in the case of temporally deterministic systems, when they will respond. This reinforces the association of control with real-time systems.

1.2 Scheduling Issues

Although we are not concerned with scheduling issues in this tutorial, a few terms should be mentioned for future reference. Real-time operating systems need to provide for either multitasking or multiprocessing (or both). In **multitasking**, the operating system must provide sufficient functionality to allow multiple programs to run on a single processor so that the illusion of simultaneity is created. This functionality includes scheduling, intertask communication and synchronization, and memory management. In **multiprocessing** operating systems, more than one processor is available to provide for simultaneity. Although multitasking may take place within any given processor, the main challenges are in process assignment, interprocessor synchronization and communication, and memory management. We will discuss some of these issues shortly, and in subsequent chapters.

There are several kinds of single processor multitasking approaches. In **round-robin systems**, each task is assigned a fixed time quantum in which to execute. A clock is used to initiate an interrupt at a rate corresponding to the time quantum. Each task executes until it completes or its time quantum expires as indicated by the clock interrupt. When a task's time quantum expires, a snapshot of the machine must be saved so that the task can be resumed later. Such schemes are used when all tasks must be equitably scheduled.

A higher priority task is said to **preempt** a lower priority task if it interrupts the lower priority task, that is, a lower priority task is running when the higher priority task signals that it is about to begin. Such schemes are used when certain processes are more critical than others. For example, in avionics systems, an imaging process may be preempted to allow a weapons control process to run. As with the round-robin system, a snapshot of the machine must be saved so that the lower priority task can be resumed when the higher priority task has finished.

Systems that use preemption schemes instead of round-robin or first-come-first-serve scheduling are called **preemptive priority systems**. The priorities assigned to each interrupt are based on the urgency of the task associated with that interrupt. Preemptive priority schemes have the associated problem of hogging of resources by higher priority tasks. In this case, the lower priority tasks are said to be facing **starvation**. There are other, non-preemptive, priority scheduling schemes, but these are of less interest to us.

Prioritized interrupts can be either fixed priority or dynamic priority. **Fixed priority systems** are less flexible in that the task priorities cannot be changed once the system is implemented. **Dynamic priority systems** can allow the priorities of tasks to change during program execution – a feature that is particularly important in threat management systems. In a special class of fixed-rate preemptive priority interrupt driven systems called **rate-monotonic systems**, priorities are assigned so that the higher the execution frequency, the higher the priority. This scheme is common in embedded applications, particularly avionics systems.

Hybrid systems include those with interrupts occurring at both fixed rates and sporadically. The sporadic interrupts may represent a critical error that requires immediate attention and thus have highest priority. This type of system is also common in embedded applications. Another type of mixed system found in commercial operating systems is a combination of round-robin and preemptive systems. Here tasks of higher priority can always preempt those of lower priority; however, if two or more tasks of the same priority are ready to run, then they run in round-robin fashion.

A concept often used as a measurement of real-time system performance is **time-loading** or **CPU utilization**, which is a measure of the percentage of non-idle processing. A system is said to be **time-overloaded** if it is 100% or more time-loaded. Time-overloading occurs in interrupt driven systems when higher priority interrupt-driven tasks execute too frequently to allow lower priority tasks to finish on time. Systems that are time-overloaded are unstable and exhibit missed deadlines and unpredictable response times.

1.3 Real-Time Design Issues

Why study real-time systems? The design and implementation of real-time systems requires the careful consideration of a variety of issues, many of which we will address in subsequent pages. Among the tasks facing the real-time system designer are:

1. Selection of hardware and software and the appropriate mix needed for a cost-effective solution.

2. The decision to take advantage of a commercial real-time operating system or to design a special operating system.

3. Prediction and measurement of CPU utilization and achieving a safe but efficient level of utilization.

4. Selection of an appropriate software language for system development.

5. Maximizing system fault tolerance and reliability through careful design and rigorous testing.

Utilization%	Zone Type	Application Types
0 – 25	overkill	various
26 – 50	very safe	various
51 – 68	safe	various
69	theoretical limit	various
70 – 99	dangerous	embedded systems
100+	overload	stressed systems

Table 1.1: CPU Utilization Zones.

6. Design and administration of tests and selection of test and development equipment.

Addressing these issues for large or even modest projects can present a staggering task.

For example, consider the evaluation of CPU utilization. Table 1.1 shows some CPU utilization ranges and subjective assessments of them. Thus, while it might be desirable to underutilize a processor for the sake of future expansion, in the near term, the additional cost of the high-powered processor may not be justified. Utilization factors in the 26%–50% range are generally considered safe – the likelihood of missing deadlines for most systems is low (yes, even at very low utilization rates, deadlines can be missed). Arbitrarily, we designate the 51%–68% range as "safe," while approximately 70% is the theoretical limit for all preemptive priority systems. Beyond 70% there is a high risk of missing deadlines, and of course CPU utilization above 100% is potentially disastrous.

Table 1.2 lists some other problem issues in real-time system design along with possible solutions and their potential drawback. We list them here simply to show the scope of the real-time design problem – in this tutorial we are concerned primarily with the prediction and reduction of CPU utilization, and optimal performance. The references list sources that discuss other issues.

1.4 What Is Real-Time Image Processing?

Real-time image processing differs from "ordinary" image processing in that the logical correctness of the system requires not only correct but also timely

Problem	Solution(s)	Possible Drawback
System modeling and design	dataflow diagrams	cannot depict control flow
Suitable programming languages	Ada	poor and unpredictable performance
Kernel selection	commercial products	poor and unpredictable performance
Intertask communication	mailboxes, queues	degrade performance
Intertask synchronization	semaphores	deadlock
Memory management	dynamic allocation	fragmentation, degraded performance
Testing	test everything	not feasible

Table 1.2: Some real-time problems, possible solutions, and potential drawbacks.

outputs; that is, semantic validity entails not only functional correctness, but also deadline satisfaction. Because of its nature, there are both supports for and obstacles to real-time image processing. On the positive side, many imaging applications are well-suited for parallelization and hence faster, parallel architectures. Furthermore, many imaging applications can be constructed without using language constructs that destroy determinism. Moreover, special real-time imaging architectures are available or can theoretically be constructed.

On the down side, many imaging applications are time critical and are computationally intensive or data intensive. And as will be discussed, there are no standard programming languages available for real-time image processing. Finally, real-time processing science itself is still struggling to produce usable results, especially for parallel processing machines. To illustrate some of these issues, we now characterize two real-time image processing systems.

Multimedia

Multimedia computing generally involves microcomputer systems equipped with high-resolution graphics, CD-ROM drives, mice, high-performance sound cards, and multitasking operating systems that support these devices. Commercial applications for multimedia computing are largely found in education, sales, and marketing.

Multimedia applications involve concurrent programs and processors, shared peripherals, and the important notion that synchronization is at least as important as timeliness. For example in multimedia applications, it is clear that audio speech output must be synchronized with the image of a person speaking (this problem, we argue, can in fact be addressed in the hardware, operating system, and language implementation). Other, possibly more difficult real-time synchronization problems exist, however. Two notable instances are video dithering and compression.

Video dithering is used to extend the available color subset by careful arrangement/rearrangement of available colors at high speed. For example, point dithering operations, which act on a pixel without regard to its neighbors, add appropriate noise to the pixel color value before it is quantized. Cluster dithering adds patterns of noise via masks to neighborhoods of pixels before quantization. In either case, if high-speed motion video is being supported, the dithering rates need to be properly synchronized with respect to lighting, video hardware performance, and visual perception cues. Work on synchronized dithering algorithms is still nonexistent.

A well-known example of dithering is the construction of composite colors from the basic red, green, blue colors (RGB) in television sets and earlier graphics displays. Here, judicious activation of the appropriate pixel colors during a preset time interval creates the perception of many more colors than just the basic three. Similarly, in black-and-white television sets, dithering has allowed for the perception of numerous shades of gray in addition to the basic white and black.

A second problem in multimedia systems involves real-time compression. For most multimedia systems, large amounts of data need to be stored and retrieved at fast rates. Often hardware boards are used, but frequently software algorithms are most cost-effective. In either case, most of the algorithms are proprietary. Of the well-known, nonproprietary algorithms, such as Huffman encoding or block truncation coding, the latter is deterministic while the former is not. An important consideration in constructing deterministic, predictable and synchronized multimedia systems is whether the proprietary compression algorithms are deterministic. In multimedia systems, we need to implement these types of compression algorithms on-the-fly and synchronously.

Virtual Reality

Virtual reality systems are complex computer simulations involving visual, audio, tactile, and other feedback to entice a person's perceptual mechanisms into believing they are actually in an artificial world. While virtual reality has obvious applications in combat simulation and training, its most promising applications are civilian, including exercise and recreation, physical rehabilitation and therapy, occupational training, and psychological diagnosis training.

For example, instead of pedaling an exercise bicycle inside an ordinary gym, a user could don a head-mounted display with head-tracking, plug in earphones, and have the impression of riding down a country road, replete with chirping birds, bumpy roads, and sun in his or her eyes. A physical therapist, wearing the same helmet-type display, could see computer-generated muscles, bones, and tendons superimposed on the body of a patient, changing with bodily movement. Construction workers learning to build skyscrapers can safely practice techniques in a virtual reality simulator. Psychologists can diagnose and/or cure patients of such phobias as vertigo, claustrophobia, and so forth, without exposing the patients to actual physical danger.

There are many other applications for virtual reality in medicine, entertainment, and so forth, and recent books, movies, and television programs have popularized this technology. These versions of the technology, however, ignore the substrate of this technology: underlying these applications are complex distributed computer control systems involving state-of-the-art electronics, software, and algorithms, and requiring sophisticated design techniques to ensure efficacy and reliability. In virtual reality, just as in multimedia, synchronization-type problems exist. For example, in virtual reality-type flight simulators, even a slight skew in the synchronization of a pilot's commands and the resultant display update (e.g., a turn is made) can cause nausea. Similarly, huge amounts of data need to be stored and retrieved quickly to simulate artificial environments. Hence real-time compression problems are of great interest to virtual reality specialists.

Chapter 2

Basic Hardware Architecture

The computer hardware has significant impact on the execution speed of a program and thus the response time of a system. In real-time systems, understanding the architecture promotes the efficient use of all resources. Although the role of current language theory is to isolate the programmer from the underlying hardware, those who have implemented practical real-time systems realize that this is impossible – if not at the design stage, certainly at the hardware/software integration stages. In this chapter we review those aspects of computer architecture that must be directly considered in the design of real-time software.

2.1 von Neumann Architecture

At the highest level, a simple computer architecture is composed of a central processing unit or CPU, a main memory, and input-output devices, as shown in Fig. 2.1. The **central processing unit** provides for arithmetic and logical operations, the main memory represents the physical address space for the CPU and is used for storage of instructions and data during program execution, and input-output (I/O) devices are used for interfacing the computer to the outside world. Note that **secondary storage** devices such as hard disks, floppy disks, tapes, and so forth, are actually I/O devices since they are not part of the physical address space of the CPU.

The process of actually retrieving, interpreting, and acting upon the instructions stored in main memory by the processor is relatively simple.

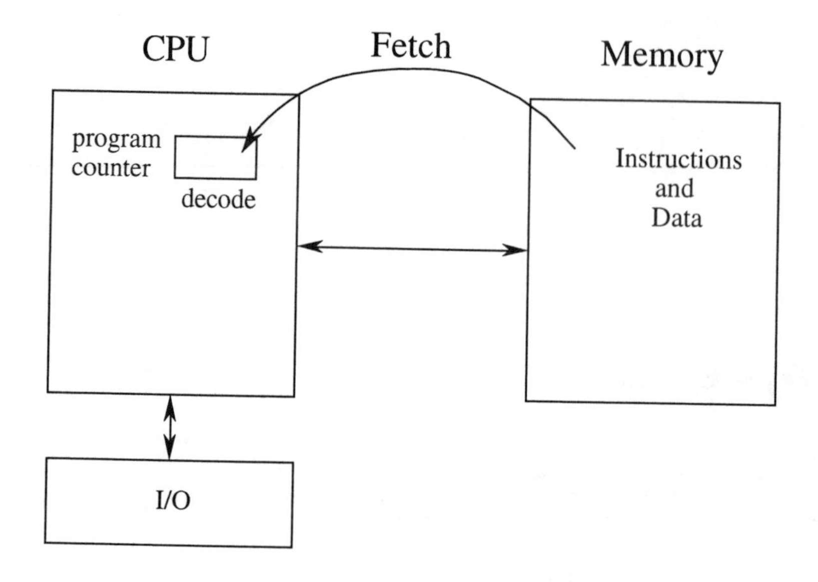

Figure 2.1: A basic computer architecture.

When power is applied, the processor begins retrieving or **fetching** the binary-coded, user-programmable instructions called **macroinstructions** from main memory via a the wires that connect them, called the **bus.*** The symbolic form of macroinstructions is known as **assembly language** or **assembly code**. Macroinstructions are usually too coarse to provide control logic for the processor, so instead are converted to a stream of more primitive instructions stored in CPU internal memory called **microinstructions** or **microcode**. A collection of microinstructions corresponding to a macroinstruction is called a **microprogram**. For example, a programmer writes the following macroinstruction:

```
Load  R1,@A
```

which has the interpretation"load register 1 with the contents of symbolic memory location A." This instruction may be decoded by the CPU into the following microprogram (comments are provided on the side):

*The bus is used to exchange memory location information ("addresses") and data between the CPU and main memory. Usually the address and data wires are referred to as the **address bus** and **data bus**, respectively.

MAR $\longleftarrow$ loc(A)	:memory address register gets the address of "A"
R/W $\longleftarrow$ 1	:set the read/write bit to 1 for a read
DST	:interrogate memory
R1 $\longleftarrow$ MDR	:register 1 gets the contents of the memory data register

where MAR, MDR, and R1 are internal CPU registers.

The task of determining which set of microinstructions corresponds to a given macroinstruction is called **decoding**. Once the microinstruction sequence has been selected, it is acted upon or **executed**. A sequence of microinstructions, once initiated, cannot be interrupted.

Serial fetch-decode-execute architectures are called **von Neumann architectures**. In standard von Neumann architectures, the serial fetch and execute process coupled with a single combined data/instruction memory, forces serial instruction and data streams. The fundamental limitation imposed by this serial process is called the **von Neumann bottleneck**. Strictly speaking, instructions or data can never be concurrently exchanged between main memory and the CPU. Moreover, because of its serial nature, the von Neumann architecture is inefficient for use with most image processing algorithms, which tend to be parallel in nature. Hence, other architectural paradigms may be needed for image processing. We discuss some of these shortly.

Modern processors are usually equipped with circuitry that enables them to handle one or more **interrupts**. An interrupt is a hardware signal that alters the sequential nature of the fetch-decode-execute cycle by transferring program control to special interrupt handler routines. A graphical illustration of an interrupt is shown in Fig. 2.2. The hardware support necessary for interrupts includes special registers such as the interrupt vector, status register, and mask register. The **interrupt vector** contains the identity of the highest priority interrupt request, the **status register** contains the value of the lowest interrupt that will presently be honored, and the **mask register** contains a bit map either enabling or disabling specific interrupts. Other specialized registers include an **interrupt register**, which contains a bit map of all pending (latched) interrupts.

Upon receipt of interrupt i, the circuitry determines if the interrupt is allowable, given the current status and mask register contents. If the interrupt is allowed, the CPU completes the current macroinstruction and then saves the program counter (PC) in interrupt return address i. The PC is then loaded with the contents of interrupt handler address i. The overall scheme is depicted in Fig. 2.3.

When the CPU is equipped to handle only one interrupt, the capability

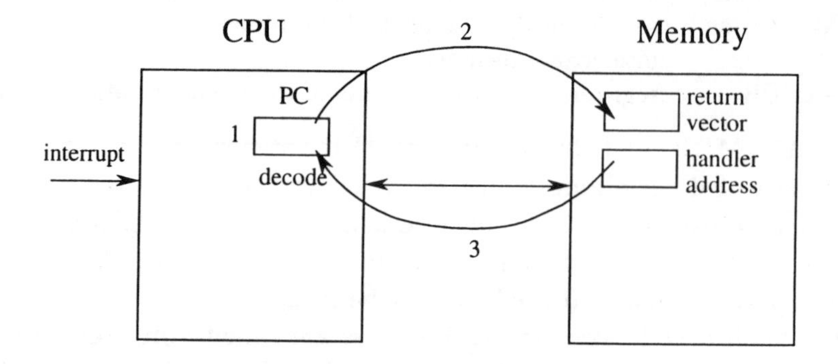

1. finish current macroinstruction
2. save PC in interrupt return location
3. load interrupt handler address
   * to return, load PC with interrupt return vector

Figure 2.2: An interrupt is a hardware signal that causes a change in program flow. Upon completion of the interrupt handler code, control is returned to the point of interruption.

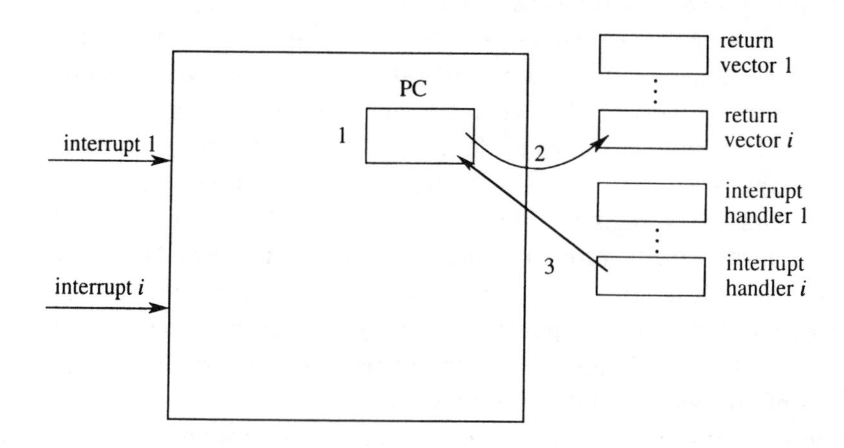

1. finish current instruction
2. save PC in return vector i
3. load interrupt handler i
   * to return load PC with return vector i

Figure 2.3: The interrupt handling process in a multiple interrupt system.

can be extended by adding an **interrupt controller** device to the system. Such a device provides the CPU with all the aforementioned features, but in either case, the CPU must be able to explicitly enable or disable interrupts via special macroinstructions. This capability is needed to facilitate the creation of the software that manages multitasking.

Because in most architectures execution of a macroinstruction cannot be interrupted, macroinstruction execution time is an important contributor to interrupt latency. The execution time of an instruction is a function of many factors, including instruction fetch and decode time, the length of its microprogram, and the number of additional memory fetches required. For example, some computer systems are microcontroller based, and because the complex macroinstruction decoding process is not required, program execution tends to be very fast.

Memory access times have a profound negative effect on real-time performance and should influence the choice of instruction modes used – both when coding in assembly language and through the careful selection of high-order language constructs. One always wishes to select or force instruction addressing modes so that the number of memory accesses is minimized.

For each macroinstruction, the number of memory fetches is influenced by the standard word length, data bus size, and the addressing mode of the instruction. The following seven addressing modes, or variants of them, are commonly found in many architectures:

- **implied mode**– involves one or more registers that are implicitly defined in the operation determined by instruction. Implied mode instructions are fast because they require only the single memory fetch for the instruction.

- **immediate mode** – involves an integer operand that is usually contained in the next address after the instruction. This type of instruction typically requires only two memory fetches, one each for the instruction and operand.

- **direct mode** – the operand is the data contained at the address specified in the address after the instruction (see Fig. 2.4). This type of instruction requires three memory fetches; one each for the instruction, data location, and data once the address is resolved.

- **indirect mode** – the operand of the instruction is a memory address containing the effective address of the address of the operand (see

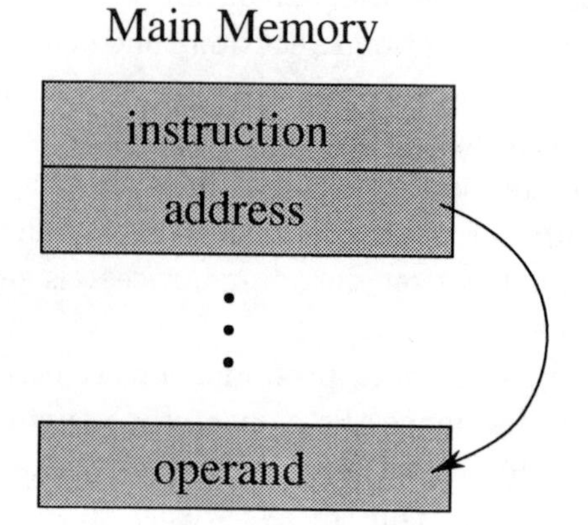

Figure 2.4: A direct mode instruction.

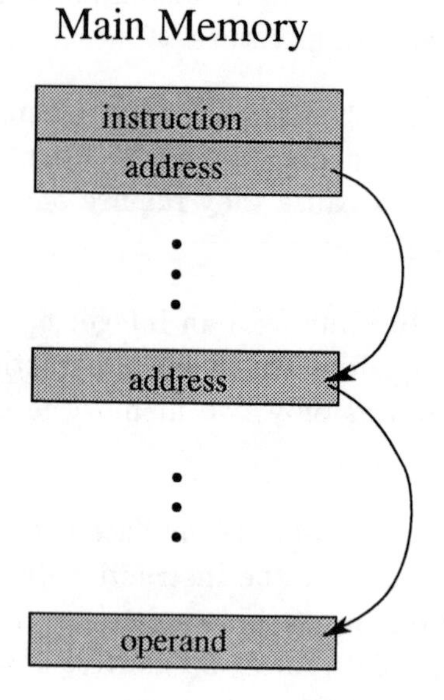

Figure 2.5: An indirect mode instruction.

Instruction Mode	Memory Accesses
implied	1
register direct	1
register indirect	2
immediate	2
direct	3
indirect	4
double indirect	5

Table 2.1: Instruction modes and the memory accesses required.

Fig. 2.5). Hence, four memory fetches are needed; one each for the instruction, operand address, contents of the operand address, and the data.

- **register direct mode** – similar to direct mode except the operand is a CPU register and not an address. This type of instruction requires one memory fetch for the instruction. The operand data are contained in registers and the target of the instruction is a register so no further memory accesses are required.
- **register indirect mode** – similar to indirect mode except the operand address is kept in a register rather than in another memory address. This type of instruction requires two memory fetches, one each for the instruction and data at the address held in the specified register.
- **double indirect mode** – is similar to an indirect mode with another level of indirection, and is frequently used to pass parameters by reference through a stack of activation records. It is also used for indexing two-dimensional arrays such as for images. Additional levels of indirection are also possible.

The number of memory access cycles for each instruction mode can be found in Table 2.1.

In real-time systems it is desirable to use instruction modes that minimize the number of memory fetches. For example, we would wish to avoid the use of indirect in favor of immediate or direct modes. This rule, however, is not inviolate since the use of indirect modes may allow for the construction of programs that are shorter and thus more memory efficient.

Input and output to a real-time system are generally managed in one of three ways: program I/O, memory-mapped I/O, or DMA. Each method has advantages and disadvantages with respect to real-time performance and ease of implementation. In **programmed I/O** special macroinstructions are used to transfer data to and from the CPU. Normally the identity of the operative CPU register is embedded in the instruction code. These instructions are CPU intensive and thus adversely impact real-time performance. Memory-mapped I/O provides a data transfer mechanism that is convenient because it does not require the use of the special macroinstructions and has the additional advantage that the CPU and other devices can share memory. In **memory-mapped I/O** reading or writing involves executing a load or store instruction on a pseudomemory address mapped to the device. In **direct memory access (DMA)**, access to the computer's memory is afforded to other devices in the system without CPU intervention, although the cooperation of a DMA controller device is required. Because CPU participation is not required, data transfer is fast. Unfortunately bus contention between the CPU and other devices can occur, sometimes causing delays in the fetch stage of the fetch-decode-execute cycle, an undesirable scenario called **cycle stealing**. Nevertheless, DMA is often the preferred method of input and output for real-time systems.

A **coprocessor** is a second, independent processor used to expand the CPU's macroinstruction set so that complicated operations need not be coded in high-level languages. Coprocessors are also used to decrease interrupt latency by reducing the slowest macroinstruction's execution time. The addition of a coprocessor to a system does not imply any form of parallelism. Instead, when the main processor encounters an instruction that is not part of its repertoire, it hands over control to the coprocessor, which presumably can execute the instruction. Coprocessors can present problems in real-time system design. For example, if the coprocessor is interruptable, then registers belonging to it should be saved when switching between tasks. This activity can increase response times. Moreover, coprocessors are only useful if the extended instruction set is appropriate for the problem at hand. For example, if the coprocessor supports morphological operations, it is essentially useless for doing Fourier transforms.

As previously discussed, memory accesses are often the most costly steps in macroinstruction execution. Modern RAM and ROM memories are largely based on semiconductor technology. RAMs are used for temporary storage for intermediate data while ROMs are used for long-term storage of programs and constants. And although current CMOS (complimentary channel

	Single Data	Multiple Data
Single Instruction	von Neumann uniprocessors RISC	Systolic Wavefront
Multiple Instruction	Pipelined machine Very long instruction word	Dataflow Transputers

Table 2.2: Classification for Computer Architectures

metal-oxide semiconductor) and NMOS (N-channel metal-oxide semiconductor) RAM provide fast access (access times around 15 nanoseconds), memory access can still be costly when performed numerous times on large images. ROM-type memories are programmable either permanently or in a way that allows reprogramming. For example, in fusible link PROM when the program is burned in, the appropriate fuses are "blown" (logical connections are removed), thereby embedding the information forever. Any other path through the device is destroyed during the burning process and so the chip is not reprogrammable. This type of ROM has low power consumption, is densely populated, has access times around 15 nanoseconds, and is readily available. Other types of ROM that are often used are EEPROM (Electrically Erasable Programmable Read Only Memory) and UVPROM (Ultra Violet Programmable Read Only Memory). These have access times around 15 nanoseconds. Memory access times are not always uniform across RAM and ROM. Hence, CPU wait states are often necessary to force a CPU to wait for a slower memory device to respond. This needs to be accounted for in CPU utilization analysis.

2.2 Architecture Classification System

Computer architectures can be classified in terms of whether there are single or multiple instruction streams and whether there are single or multiple data streams, as shown in Table 2.2.

Most computer systems qualify as **single instruction stream, single data stream (SISD)** systems. An SISD computer has a central processor unit that processes a single instruction at a time and a single piece of data at a time. In most SISD machines there is synchronized communication between the CPU and memory (MEM). Instructions to be executed by the CPU are stored in memory, along with data. The CPU specifies addresses of mem-

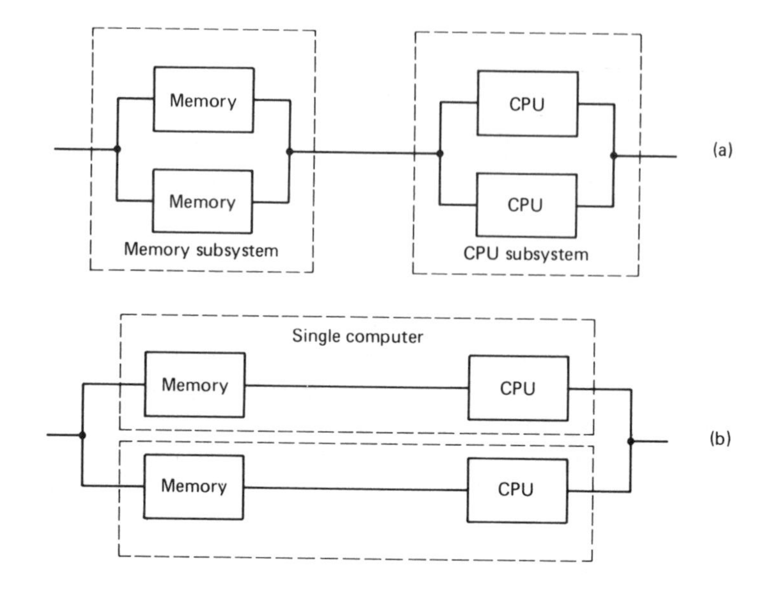

Figure 2.6: Parallel CPU systems.

ory that contain instructions required for processing data. For a sequence of instructions to be carried out in the proper order, the computer must fetch and execute each instruction in turn. In an SISD system, instructions are executed sequentially and each instruction operates on a single datum drawn from memory. The system most likely will include an I/O processor (IOP); however, if the IOP does not possess independent computational facilities, then it does not enhance the speed of actual data processing, so the system remains functionally an SISD system. To this point we have been mainly discussing SISD von Neumann systems.

Faster processing can be achieved by concurrent processing of more than a single instruction, more than a single datum, or both. One way of achieving multiprocessing is to organize two or more CPUs in conjunction with two or more memory modules, as in Fig. 2.6. For the system in part (a), each CPU has access to either memory; for the system in part (b), there are two distinct CPU-MEM systems. In either case, concurrent execution results from two SISD CPUs functioning in parallel. Sequentiality is not abandoned; rather, several CPUs are joined, augmented by the utilization of multiple memory modules. Since the main program must continue to run according to the predetermined control sequence, a complex switching network is necessary to route instructions and data between distinct processors and memory modules. As the number of processors increases, so does the complexity of both the switching network and the data paths. As a consequence, memory latency occurs: processors remain idle while they await responses to requests for data. The operating system, which might be located in one location within the system or spread throughout the system, must be able to accommodate the parallel nature of the system. Finally, programs must be written in order to take advantage of the parallelism. The programmer, or a compiler, must detect independent data flows in the algorithm, write these to the independent CPUs while at the same time being careful that data at an address called by one CPU has not already been altered by a different CPU, and merge the parallel flows at appropriate times. In the next sections, we describe these architectures in some detail. Later, we will show how many of these architectures can be employed to improve the real-time performance of compatible imaging algorithms.

2.3 SISD Processors

We have already discussed the serial input stream, serial data stream architectures by introducing the von Neumann architecture. Since its acceptance as the standard processing paradigm, variations have been introduced in an attempt to overcome the von Neumann bottleneck. We discuss these now.

RISC (reduced instruction set computer) architectures are a special class of SISD architectures. In RISC architectures, a limited number of macroinstruction types and addressing modes simplifies the decode and macroinstruction execution process, hence improving overall execution times. While many RISC architectures include pipelining, and hence become multiple instruction stream, pipelining, which is discussed subsequently, is not a requisite characteristic of RISC.

RISC processors also present certain advantages in real-time processing since the maximum macroinstruction execution time (and thus interrupt latency) is less than that for a CISC machine. RISC also uses a high percentage of fast register direct and register indirect instructions, which can greatly reduce interrupt latency.

Complex instruction set computers (CISC) are the polar opposite of RISC machines. CISC processors are characterized by a large number of complex instructions involving long microprograms, numerous multilevel addressing modes, and sophisticated CPUs. Until the mid-1970s, most mainframe and minicomputers were designed as CISC processors. Current opinion, however, is that "less is more" and new development leans away from CISC processing toward RISC.

Because one component of instruction execution time is memory access, one way of increasing performance is to reduce access times. **Memory caching** is a technique in which frequently used segments of main memory are stored in a faster bank of memory that is local to the CPU (called a **cache**). The cache consists of high-speed memory that is loaded with blocks of main memory according to various rules (the cache is necessarily much smaller than main memory). If an instruction or datum is requested from main memory, and it resides in the cache, memory access time will be reduced. If the instruction or datum is not found in the cache, the cache must be reloaded with a block of contiguous main memory that contains the requested address or datum, and the memory access time for the request will actually be increased. Note that any changes to data in the cache must eventually be reflected in the corresponding part of main memory. The percentage of time in which the requested instruction or datum is actually in the cache is called the **cache hit ratio**. A high hit ratio is of course desired. Code where instructions tend to be accessed linearly (as opposed to "spaghetti" code that jumps around) will have a high **locality-of-reference** and tend to have a high hit ratio. Fortunately, imaging applications tend to have a high locality-of-reference because of the frequent use of looping constructs, although manipulating large images that do not fit in cache can lower the hit ratio dramatically.

Because the cache renders access times different depending on the sequence of execution (which is data dependent), performance prediction is practically nondeterministic. Strictly speaking, deterministic performance can be had by always assuming that at all necessary points, cache must be reloaded (a worst-case scenario). However, this is a pathological condition and, in well-written code, average case performance is almost always improved by using a cache.

Microcontrollers are a type of von Neumann architecture where there is no decoding of macroinstructions. Rather, software is programmed directly in the low-level microcode, allowing for very rapid execution of these primitive instructions. Microcontrollers are used in many imaging hardware devices (such as digitizing cameras, frame grabbers). However, microcontrollers are not used as general-purpose image processors.

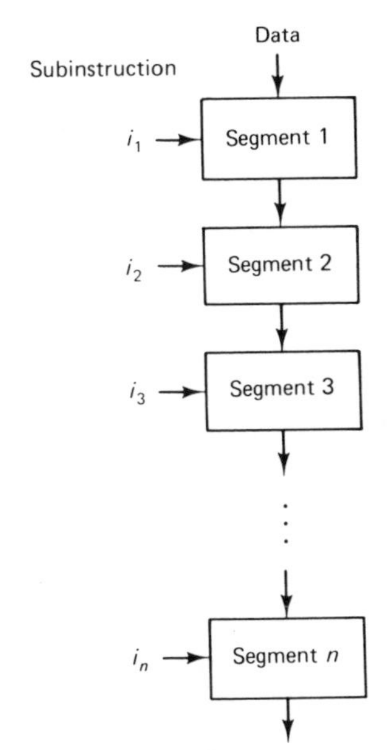

Figure 2.7: Pipeline segments.

2.4 MISD Processors

A machine that can process two or more instructions concurrently on a single datum is called a **multiple instruction, single data** (MISD) machine. We now review several such architectures.

A rudimentary form of instruction concurrency is achieved by **pipelining**. Architecturally, pipelines require disjoint hardware to facilitate concurrent operations. Each disjoint processing circuit is called a **segment** or **stage**. Each segment is like a stage in an assembly line, with partial computation being performed concurrently with computations at other stages. At each stage, a distinctive operation is carried out and the final result is obtained after all stages have been visited (see Fig. 2.7). Concurrent computation is facilitated by associating a register with each segment in the pipeline, so that, relative to data, the segments are isolated and can concurrently process different data.

Pipelining necessitates decomposition of macroinstructions into sequences of more primitive operations. These primitive operations are employed in a concurrent manner by overlapping steps in the instruction cycle. Complicated instructions may often require the same resource several times (a segment may need to be visited more than once). Consequently, pipeline

Clock Cycle:	0	1	2	3	4	5	6	7
Instruction 1	IF	ID	EX	MA	WR			
Instruction 2		IF	ID	EX	MA	WR		
Instruction 3			IF	ID	EX	MA	WR	
Instruction 4				IF	ID	EX	MA	WR

Table 2.3: Timing diagram for a five-stage pipeline.

scheduling difficulties may exist, and collisions may occur if inadequate scheduling is performed.

As an example, consider the following five-stage pipeline consisting of the following stages:

IF Instruction fetch – retrieve instruction from memory

ID Instruction decode – select mode and start address of microprogram

EX Execute – begin executing microprogram

MA Memory access – a datum is loaded from or stored to memory

WR Write back – cache is updated accordingly.

A timing diagram showing how the instructions overlap is shown in Table 2.3. Note that if one of the instructions in the pipeline is a branch instruction, then upon execution, the prefetched instructions in the pipeline become obsolete and must be purged or **flushed** from the pipeline. As this is a setback, subsequent instructions will take longer to execute until the pipeline becomes full again. Because the prediction of when the pipeline is flushed is practically impossible, performance upper bounds in pipelined architectures can really only be based on worst case performance (that is, the pipeline has to be flushed whenever possible). Again this illustrates the trade-off between improving average case performance at the expense of worst case performance.

Since in a serial instruction stream, the von Neumann bottleneck limits the rate of instruction execution, one work-around is to make those instructions more meaningful, so that more functionality is packed into each instruction. Such computers tend to be implemented with microinstructions that have very long bit-lengths (and can control more internal CPU signals).

Hence, rather than breaking down macroinstructions into numerous microinstructions, several (nonconflicting) macroinstructions can be combined into a few microinstructions. For example, if object code was generated that called for a load register followed by an increment of another register, these two instructions could be executed simultaneously by the processor (or at least appear so at the macroinstruction level) with a series of long microinstructions. Since only nonconflicting instructions can be combined, any two accessing the bus conflict. Hence, only one instruction can access the bus and so the very-long-instruction-word computer is MISD.

2.5 SIMD Processors

In **single instruction stream, multiple data stream (SIMD)** computers each processing element is executing the same (and only) instruction, but on different data. Two types of SIMD processors are systolic processors and wavefront processors.

A **systolic system** consists of a set of interconnected cells, each capable of performing a simple operation. Information flows in a pipelined fashion and outside communication occurs only at boundary cells. Systolic architectures are designed for **compute-bound** computations in which the number of operations is large in comparison to the number of I/O instructions (I/O is generally performed by other processors). Data are used at the various cells and are pumped from cell to cell in the architecture. A systolic array usually has a simple geometric structure involving local connections, each cell performing an identical task. The synchronous pulsing of data is controlled by an external clock or "heartbeat," hence the term "systolic."

A **wavefront array processor** is similar to a systolic processor except that there is no external clock and array processors operate asynchronously. Results are transmitted when computed and computing commences whenever all operands are available. Thus, wavefront processors are a cross between systolic and dataflow processors (to be discussed shortly).

2.6 MIMD Processors

Multiple instruction stream, multiple data stream (MIMD) computers include dataflow processors and transputers. Each is characterized by a large number of processing elements, each capable of executing numerous instructions.

In **dataflow architectures**, control flow is determined by the availability of data (operands), rather than by any specific control apparatus within the program. Whereas traditional architectures are sequential, control being maintained by a program counter that causes processing to follow a user-determined path (or flow chart), dataflow design eliminates sequentiality and allows the stream of computation to proceed along a program graph without regard to any prerequisite control (for this reason traditional dataflow diagrams are often used to design algorithms for dataflow machines).

The principle is straightforward: if the operands for a specific operation are available and the destination of the result is prepared to receive it, then there is no need to await the processing of other operands elsewhere within the program. Dataflow processors execute their instructions as soon as all required operands have arrived. After execution (firing) of an instruction, the result is transmitted to all other instructions that require it as an input. Counters are employed for each instruction to keep count of the number of operands that have arrived, as well as how many more are needed for instruction execution. There is no overall program counter to synchronize instruction execution; instructions are executed in an asynchronous manner. One consequence of the flow-oriented design is that dataflow machines utilize graphical languages that are similar to directed graphs. Logical entities called **tokens** are employed to represent the dynamics of a dataflow system. During actual processing, tokens are often held in queues.

A dataflow operator with n variables, which is represented as a node on the program graph, can be in one of several states. These are illustrated in Fig. 2.8, where a dot is used to denote the presence of a token. Part (a) of the figure illustrates the **waiting state**. Here, the function cannot be executed since not all input lines contain a token. Part (b) depicts the **ready**, or **enabled state**, where all necessary tokens have arrived and the input lines are full. Part (c) illustrates an activated dataflow function firing, and part (d) shows a dataflow operator outputting the result generated.

It is possible that several tokens might be present on an input line to a dataflow processor. This situation is illustrated in Fig. 2.9, where no feedback mechanism is employed: execution proceeds on a first-come-first-served basis.

Information in a dataflow system is often organized utilizing **activity packets** (or **templates**). The opcode of the operation is specified at the top of the activity template (Fig. 2.10). The number of operands required for execution of the instruction is also specified in the template. As operands arrive, this number is decremented in a counter until it reaches zero, at which

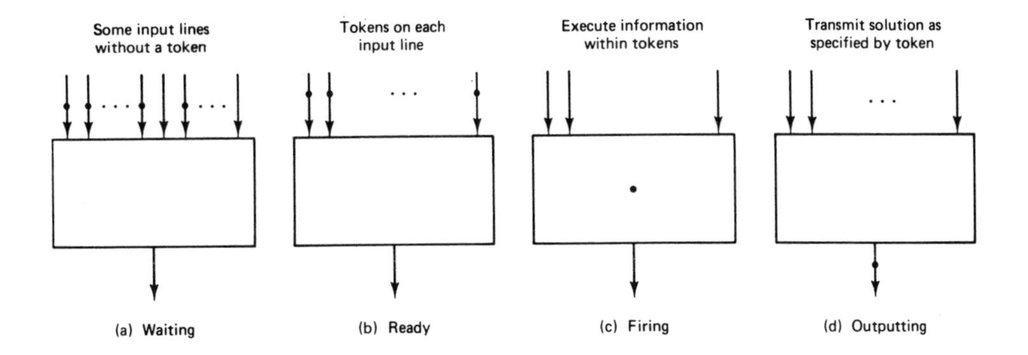

Figure 2.8: States in a dataflow machine.

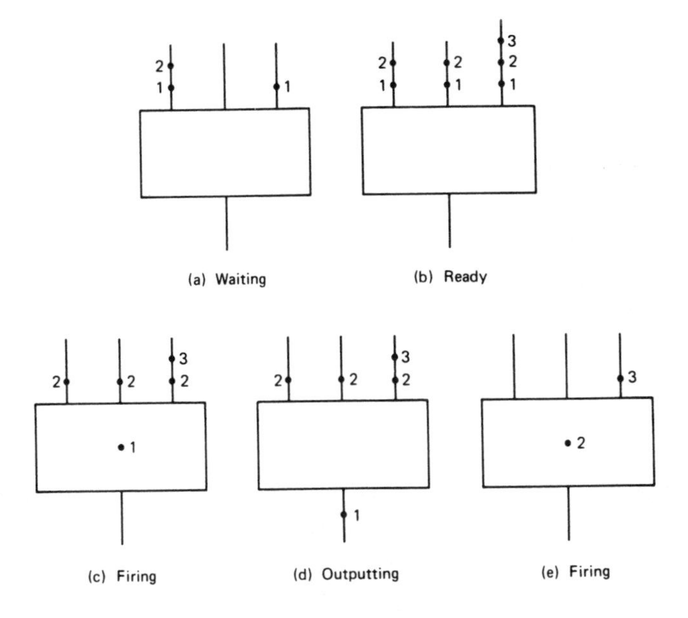

Figure 2.9: Tokens on input lines to dataflow processors.

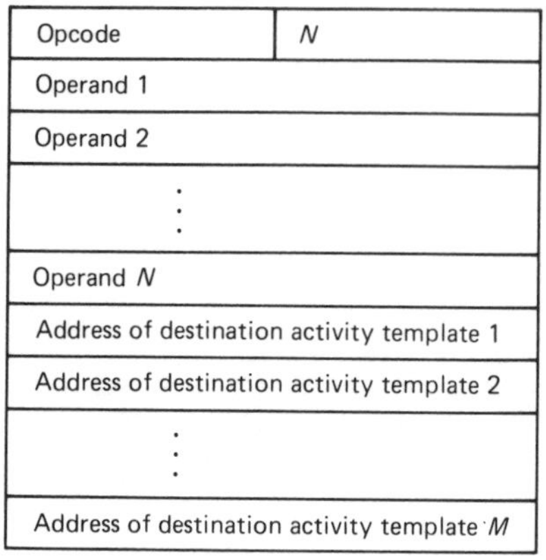

Opcode	N
Operand 1	
Operand 2	
⋮	
Operand N	
Address of destination activity template 1	
Address of destination activity template 2	
⋮	
Address of destination activity template M	

Figure 2.10: Activity packet.

point execution of the instruction occurs. There is room in the template to store each of the operands as they arrive, as well as the addresses of all other activity templates that need the result of the operation. As soon as an operation fires, the result is transmitted to all addresses specified and stored as an operand in each destination activity template.

The major drawback to dataflow architectures is the lack of support tools such as programming languages (which must be graphical), debuggers, performance analyzers, and so forth. Moreover, performance analysis is sketchy at best, unless branching is prohibited.

Transputers are fully self-sufficient, multiple instruction set von Neumann processors. The instruction set includes directives to send data or receive data via ports that are connected to other transputers. Transputers, while capable of acting as uniprocessors, are best utilized when connected in a mesh configuration. In a sense, the transputer provides a wavefront or systolic processing capability, but without the restriction of a single instruction. Indeed, by providing each transputer mesh with an appropriate stream of data and synchronization signals, reconfigurable wavefront or systolic computers can be implemented.

Chapter 3

Linear Image Processing Algorithms

Real-time processing for digital imaging concerns efficient, deterministic implementation of algorithms whose inputs include digital images and whose outputs are digital images, numerical features, symbolic representations, or decisions. Computation bottlenecks appear in many forms and, to some extent, each requires its own real-time implementation via algorithm design, software, hardware, or a combination thereof. For example, the von Neumann bottleneck prevents the parallelization of certain algorithms on serial processors. Rather than present a myriad of individual approaches to specific real-time imaging problems, we shall focus on a small number of computation tasks that occur across various application domains.

The present chapter introduces basic definitions, applications, and computations pertaining to linear processing, in particular, windowed convolution. The next two chapters discuss matrix transformations and nonlinear processing. The imaging operators discussed in Chapters 3, 4, and 5 provide a class of processing problems to which we can subsequently refer when we discuss software and hardware solutions to real-time computation.

3.1 Convolution

A **digital image** is a function of two discrete variables. If an image is denoted by f, then at each pixel (i, j) its gray-value is denoted by $f(i, j)$. We conceive of an image as having infinite extent across the Cartesian grid; however, in practice, it is limited to some frame, so that i and j are bounded.

Similarly, we may conceive of the gray values of f being unlimited, taking on any integer value, but in practice the gray range is also bounded and is typically assumed to be between 0 and 2^{n-1}, inclusive, for some integer $n > 0$. A typically sized practical image might be defined over a 256×256 frame and have gray values between 0 and 255. Software implementations of images usually take the form of arrays or linked lists, or more complicated data structures such as quad-trees (discussed in Chapter 4). Mathematical operations performed on images might yield gray values outside the specified range. Such computations pose no problem in and of themselves, but they might produce wraparound problems owing to bit limits set within the computer and there might need to be some truncation of values at the end of the algorithm, for instance, to display the image. Moreover, numerical operations on images often produce noninteger values and these require quantization.

Perhaps the most basic imaging computation is **windowed convolution**. For convolution, there is a mask g of numerical weights defined over some window W, usually square, the window is translated across a digital image pixel by pixel, and at each pixel the arithmetic sum of products between the mask weights and the corresponding image pixels in the translated window is taken. Given a $(2m+1) \times (2m+1)$ mask g of weights $g(i,j)$, i, $j = -m, -m+1, \ldots, m$, where for notational ease we have restricted ourselves to an odd-sized window, and an image $f = f(i,j)$, the **convolution** image $h = g * f$ is defined at pixel (i,j) by

$$h(i,j) = \sum_{r=-m}^{m} \sum_{s=-m}^{m} g(r,s) f(i+r, j+s). \tag{3.1}$$

The convolution value $h(i,j)$ is computed by centering the window W at (i,j). The set of pixels covered when the window is placed at (i,j) is denoted by $W_{(i,j)}$, where the subscript (i,j) is used to indicate that the set W has been translated to (i,j). Convolution is a linear operator and is spatially invariant because it operates in the same manner at each pixel.

Since, in principle, an image is defined over the entire Cartesian grid, $h = g * f$, as given in Eq. 3.1, is well-defined; that is, the double sum can be computed for all window positions over the image. Since, in practice, the image exists within a frame, the question arises as to how to define $h(i,j)$ if $W_{(i,j)}$ extends outside the frame. This problem of windowing near the frame boundary arises regularly and there are various ways of handling it. For linear processing, these include defining the image to be zero outside its frame, extending the boundary values outward into the complement of the frame, and extending the image periodically. Since our interest is computation, we shall concentrate on algorithmic implementation of the basic operational definitions and leave boundary nuances to a text on image processing. Insofar as linear convolution is concerned, when necessary we shall take the mathematically straightforward approach of defining the image to be zero outside its frame. This decision yields significant real-time performance benefits, in that it generally obviates calculations involving pixels that are outside the image frame.

As defined in Eq. 3.1, discrete convolution represents a digital version of convolution for real-valued functions defined over the real line. In the analog setting,

$$h(x) = \int_{-\infty}^{\infty} g(t)f(x+t)dt. \tag{3.2}$$

We shall employ one-dimensional examples to illustrate some of the effects of convolution.

In Eqs. 3.1 and 3.2 the convolution sum and integral are defined with $f(i+r,j+s)$ and $f(x+t)$, respectively. Usually these are defined with $f(i-r,j-s)$ and $f(x-t)$, respectively. Mathematically, the choice is unimportant; however, we have chosen to use $f(i+r,j+s)$ since, geometrically, it maintains the mask orientation when the convolution sum is computed.

3.2 Linear Noise Suppression

In Eq. 3.1, there are $(2m+1)^2$ pixels in the window over which g is defined. If all weights are equal and sum to one, then $g(r,s) = (2m+1)^{-2}$ for all r and s and Eq. 3.1 defines the average value of the image f over the pixels in the window when it is centered at pixel (i,j). More generally, if $g(r,s) \geq 0$ for all (r,s) and the sum of the values $g(r,s)$ is one, then $h(i,j)$ is a weighted average of the image values in the window centered at (i,j). Consequently, $h = g * f$ is often called a **moving average**. We shall use moving average terminology only when the latter weighting conditions are satisfied.

A moving average acts as a **smoothing filter** and can be used to suppress random additive pixel noise. To see this, suppose f is the ideal image, n is the noise image, and the observed noisy image is $f+n$. Since convolution is a linear operator, convolution with mask g yields

$$g*(f+n) = g*f + g*n. \tag{3.3}$$

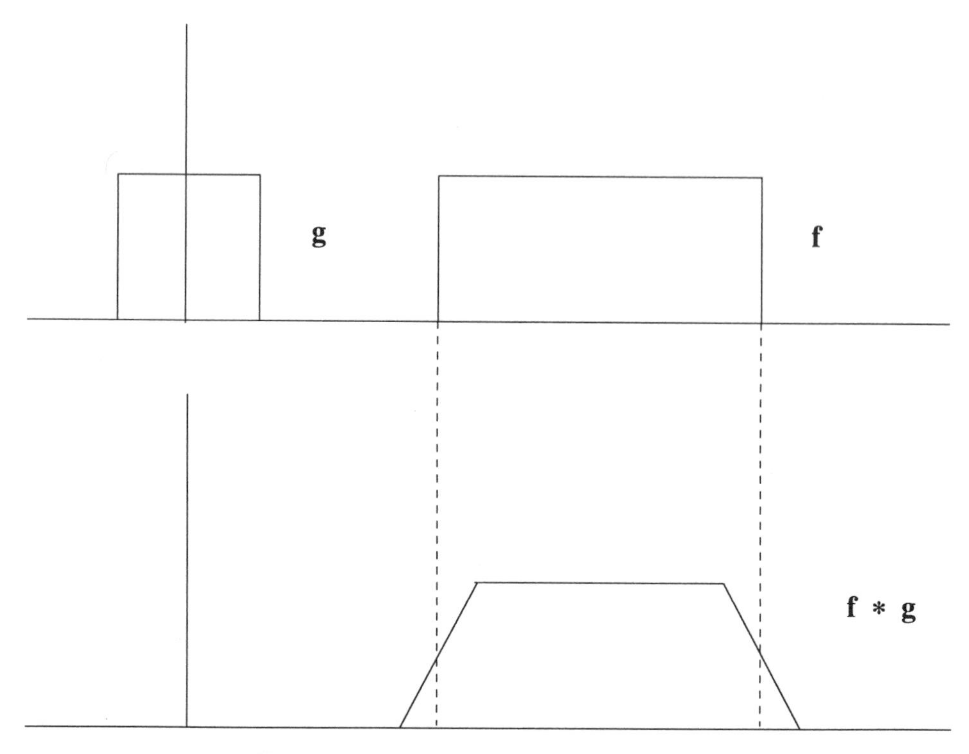

Figure 3.1: Edge effect of smoothing.

Suppose g is the mask with equal weights. If the noise is random with mean zero, then (in a probabilistic sense) $g * n$ is close to zero, so that the noise is suppressed in the output image. The down side is that the convolved image $g * f$ is a smoothed version of f, so that key information, such as edges, might be lost or made less distinctive (see Fig. 3.1 for a one-dimensional illustration of edge flattening).

The problem can be mitigated by using varying weights in the mask that accentuate the center pixel and those near to it. There is a trade-off here. Such weighting makes $g * f$ closer to f, but at the cost of not suppressing the noise as well. Choosing the best weights is the basic problem of optimal linear filtering. Here we only present some examples to show the effect of noise suppression via linear filtering. Figures 3.2 through 3.5 show an original image, a salt-and-pepper corrupted version of the image, and four images obtained from the noisy image via a 5×5 mean, a 5×5 center-weighted moving average, a 3×3 mean, and a 3×3 center-weighted moving average.

(a)

(b)

Figure 3.2: Moving-average restoration of impulse-degraded aerial scene: (a) original image; (b) degraded image.

(c)

(d)

Figure 3.2: Moving-average restoration of impulse-degraded aerial scene: (c) 5×5 mean; (d) 5×5 center-weighted.

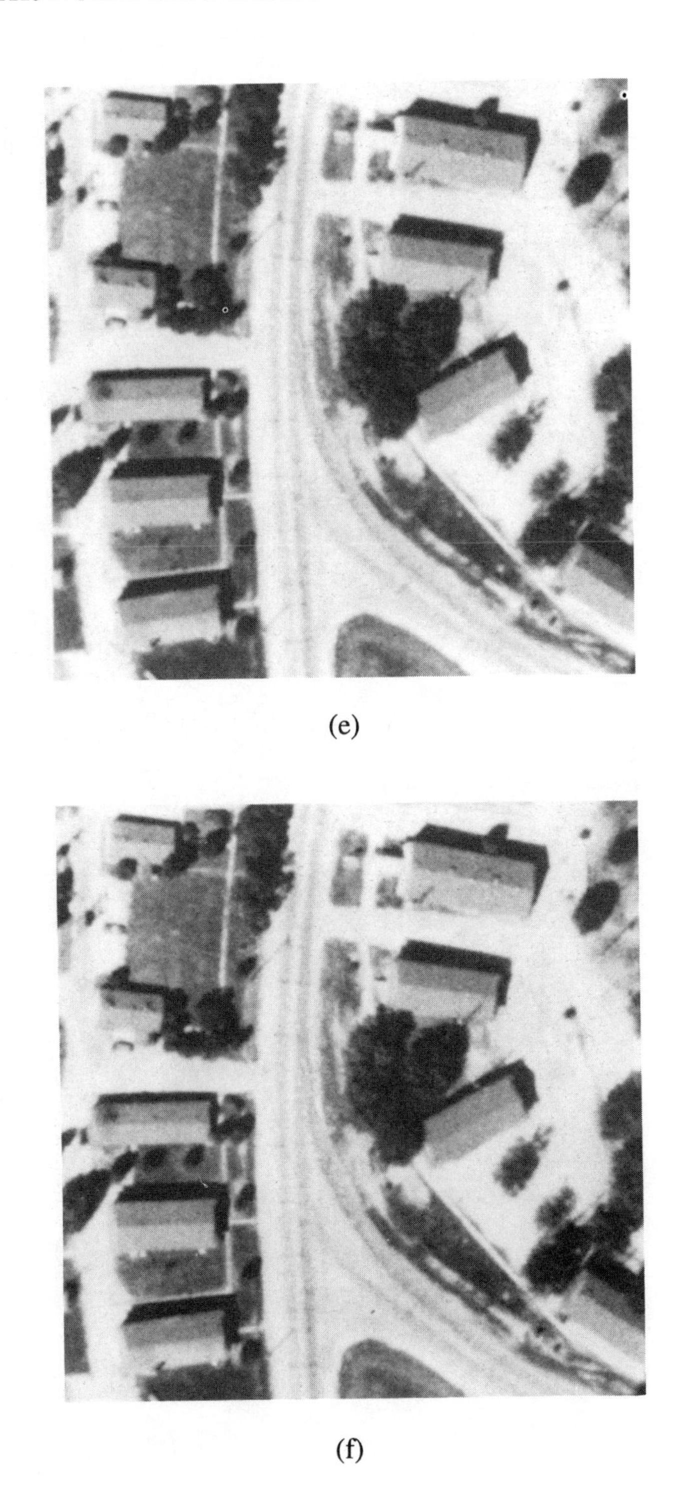

(e)

(f)

Figure 3.2: Moving-average restoration of impulse-degraded aerial scene: (e) 3×3 mean (f) 3×3 center-weighted.

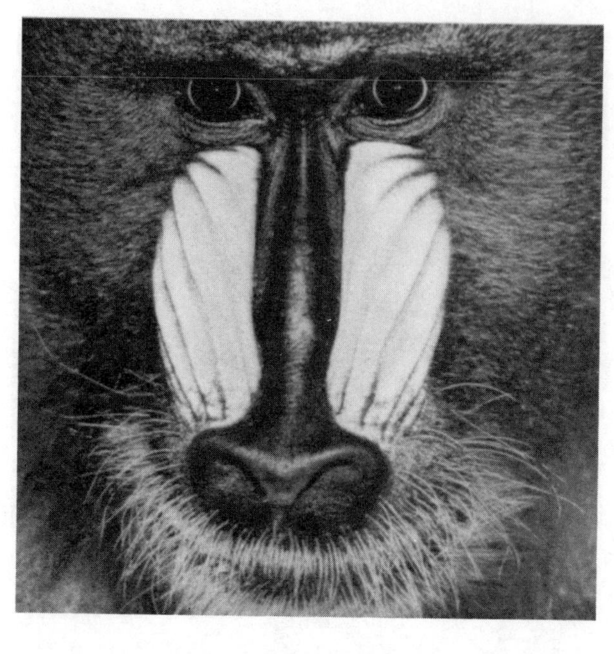

(a)

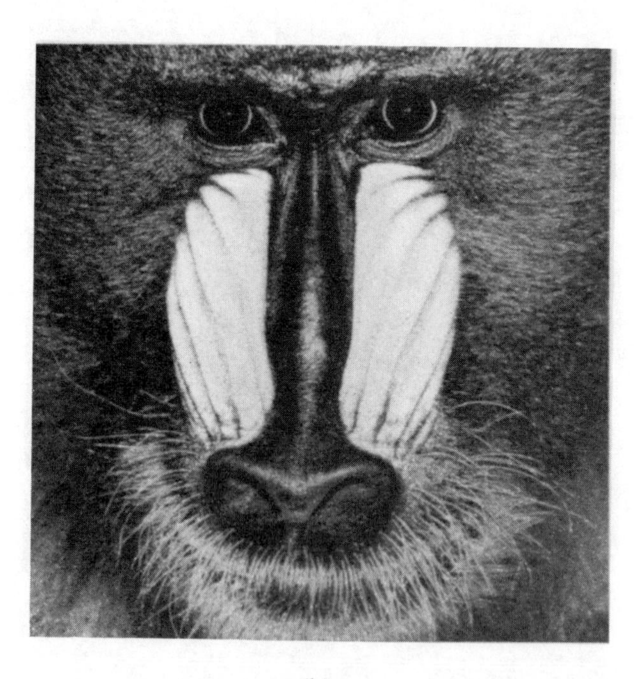

(b)

Figure 3.3: Moving-average restoration of impulse-degraded mandrill: (a) original image; (b) degraded image.

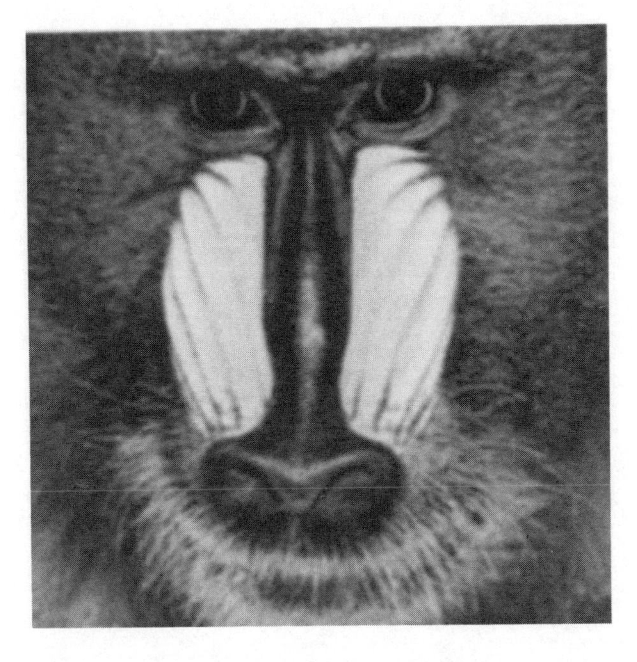

(c)

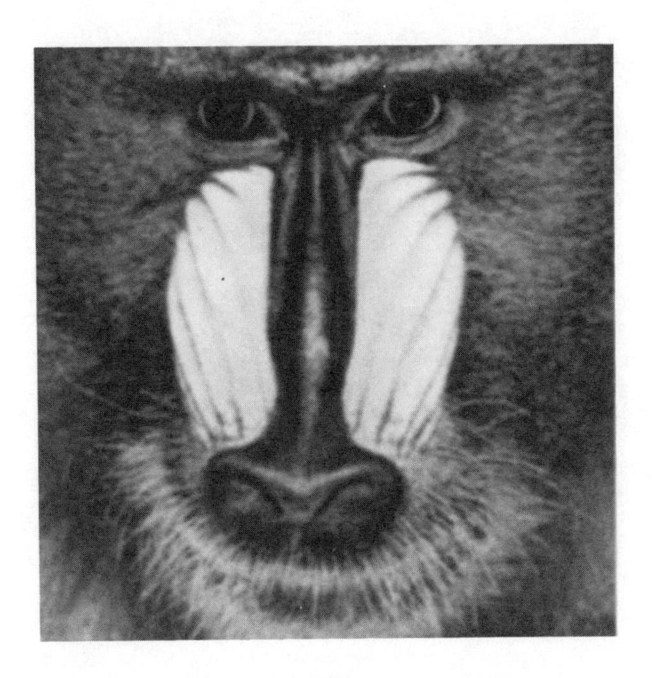

(d)

Figure 3.3: Moving-average restoration of impulse-degraded mandrill: (c) 5×5 mean; (d) 5×5 center-weighted.

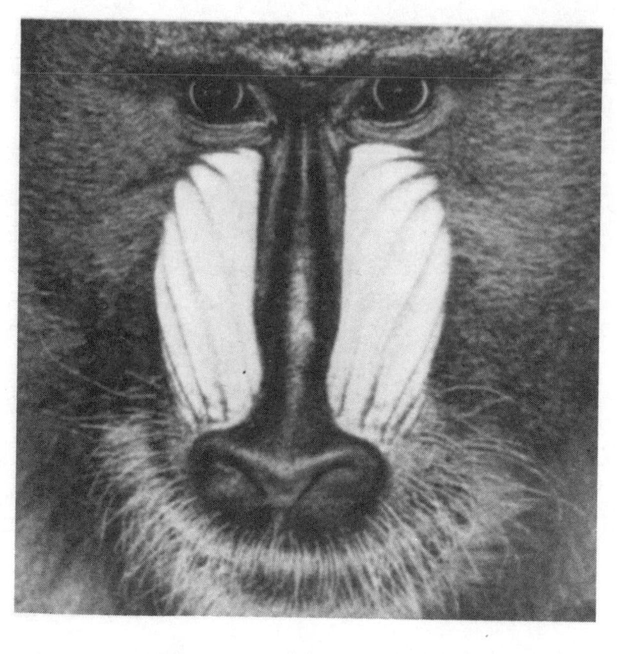

(e)

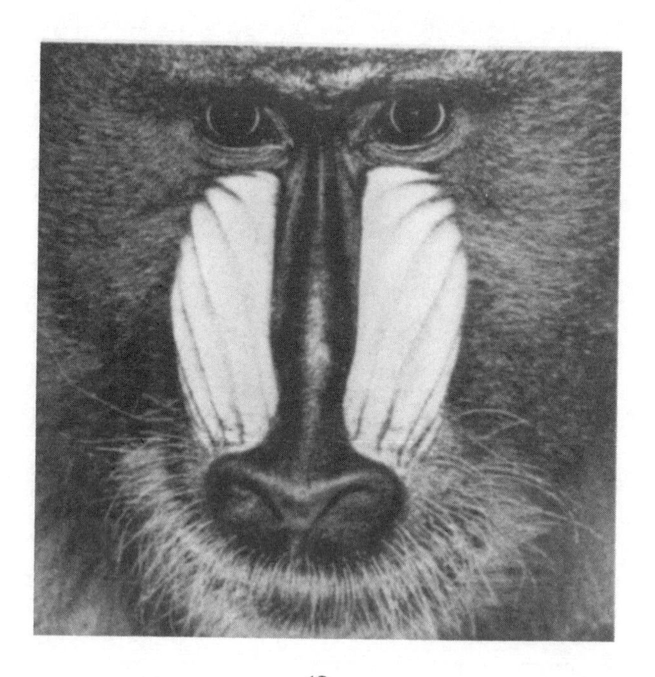

(f)

Figure 3.3: Moving-average restoration of impulse-degraded mandrill: (e) 3×3 mean; (f) 3×3 center-weighted.

(a)

(b)

Figure 3.4: Moving-average restoration of impulse-degraded bridge scene: (a) original image; (b) degraded image.

(c)

(d)

Figure 3.4: Moving-average restoration of impulse-degraded bridge scene: (c) 5×5 mean; (d) 5×5 center-weighted.

(e)

(f)

Figure 3.4: Moving-average restoration of impulse-degraded bridge scene: (e) 3×3 mean; (f) 3×3 center-weighted.

(a)

(b)

Figure 3.5: Moving-average restoration of impulse-degraded house scene: (a) original image; (b) degraded image.

(c)

(d)

Figure 3.5: Moving-average restoration of impulse-degraded house scene: (c) 5×5 mean; (d) 5×5 center-weighted.

(e)

(f)

Figure 3.5: Moving-average restoration of impulse-degraded house scene: (e) 3×3 mean; (f) 3×3 center-weighted.

3.3 Linear Edge Detection

The real-time detection of edges in digital images is widely used in robot navigation, industrial inspection, and virtual reality. Convolution is often used to provide an **edge filter** that takes in a gray-scale image and yields a binary image whose 1-valued pixels are meant to represent an edge within the original image. In essence, convolution provides gradient operators in both the horizontal and vertical directions and the final edge image is formed by marking pixels of high gradient intensity. The general scheme for edge detection is given by the block diagram in Fig. 3.6. The image f is convolution filtered by horizontal and vertical gradient masks, a pixelwise norm (magnitude operator) is applied to the resulting directional gradient images to produce an overall gradient image, and the norm image is then thresholded to provide a binary edge image. Under the assumption that a pixel at which there is a high gradient has a high probability of being an edge pixel, the intent of the procedure is clear: mark pixels of high gradient.

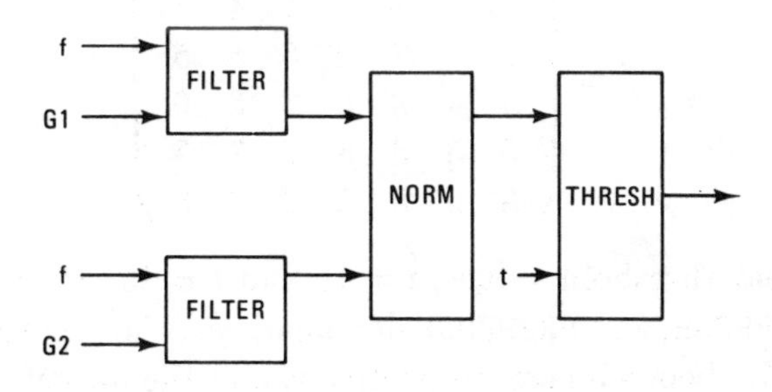

Figure 3.6: Gradient edge detection.

To illustrate the procedure, consider the **Prewitt gradient masks**

$$g_x = \begin{pmatrix} -1 & 0 & 1 \\ -1 & \mathbf{0} & 1 \\ -1 & 0 & 1 \end{pmatrix} \quad g_y = \begin{pmatrix} 1 & 1 & 1 \\ 0 & \mathbf{0} & 0 \\ -1 & -1 & -1 \end{pmatrix} \tag{3.4}$$

(bold font indicates the origin). The convolutions $f * g_x$ and $f * g_y$ are gradient images in the x and y directions, respectively. A common pixelwise norm is the maximum of the absolute values. For the maximum norm, at pixel (i,j) the output of the norm block of Fig. 3.6 is

$$k(i,j) = \max\{|\,(f * g_x)(i,j)\,|, |\,(f * g_y)(i,j)\,|\}. \tag{3.5}$$

Another common pixelwise norm is the sum of the absolute values. This norm is called the 1-norm and for it, at pixel (i,j) the output of the norm block is

$$k(i,j) = |\,(f * g_x)(i,j)\,| + |\,(f * g_y)(i,j)\,|. \tag{3.6}$$

The final edge output is the threshold image $T[k;t]$ defined by

$$T[k;t](i,j) = \begin{cases} 1, & \text{if } k(i,j) \geq t \\ 0, & \text{if } k(i,j) < t, \end{cases} \tag{3.7}$$

where t is the threshold value. The assumption, which may or may not be correct, is that (i,j) is an edge pixel if and only if $k(i,j) \geq t$. Obviously, selection of t is crucial. A walk–through of gradient edge detection using the Prewitt masks and the 1-norm is given in Fig. 3.7 for the image

$$f = \begin{pmatrix} 2 & 1 & 1 & 1 & 0 & 1 & 4 \\ 2 & 1 & 1 & 2 & 1 & 4 & 4 \\ 2 & 3 & 2 & 2 & 5 & 5 & 5 \\ 4 & 4 & 3 & 6 & 7 & 6 & 6 \\ 5 & 5 & 8 & 8 & 7 & 7 & 7 \\ 6 & 9 & 8 & 8 & 8 & 9 & 8 \end{pmatrix} \tag{3.8}$$

and two different threshold values, $t = 10$ and $t = 15$. In the figure, PREW1, PREW2, MAG1, THRESH, and PREWEDGE denote g_x, g_y, the 1-norm, the threshold operator, and final edge image, respectively, and the dashes denote 0s in the edge images. Note that frame-boundary effects have been circumvented by not processing the boundary pixels in the input image. Once again, this provides a real-time performance benefit.

Other masks besides those of Eq. 3.4 can be used. A popular mask set consists of the **Sobel masks:**

$$g_x = \begin{pmatrix} -1 & 0 & 1 \\ -2 & \mathbf{0} & 2 \\ -1 & 0 & 1 \end{pmatrix} \quad g_y = \begin{pmatrix} 1 & 2 & 1 \\ 0 & \mathbf{0} & 0 \\ -1 & -2 & -1 \end{pmatrix}. \tag{3.9}$$

Whereas the Prewitt masks weight the pixels above and below, and before and after, equally, the Sobel masks give greater weight to the pixels on the horizontal for g_x and on the vertical for g_y.

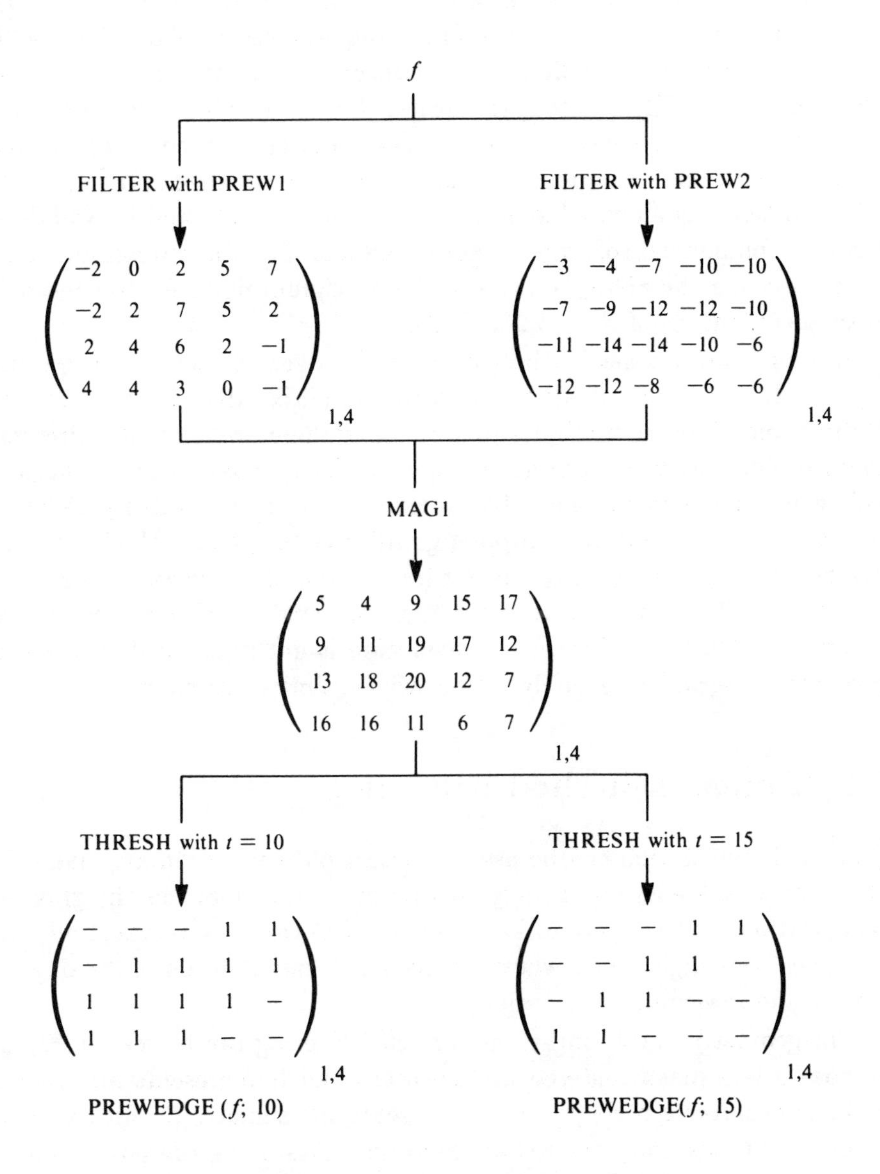

Figure 3.7: Walk-through of gradient edge detection.

Gradient edge detection is susceptible to noise-induced false detection of edge pixels. Selection of an appropriate threshold is also problematic: too low a threshold gives wide edges and too many edges; too high a threshold can result in missing key edge information, for instance, by yielding broken or partial edges. Prefiltering to reduce noise is often necessary, but a linear smoothing filter flattens gradients, thereby making edges both wider and harder to detect by thresholding. (In Chapter 5 we consider nonlinear alternatives to noise suppression that better preserve edges.) Even with all of its problems, gradient edge detection remains popular and is available on standard commercial software. Figures 3.8 and 3.9 illustrate gradient edge detection using the Sobel masks and the maximum norm for the aerial and house scene images of Figs. 3.2 and 3.5.

Rather than use masks that only detect edges in the $x-$ and $y-$ directions, a slightly different approach is to use masks that detect changes in all directions that are multiples of 45°. Since there are eight 45° directions, the procedure involves filtering by eight different masks, each being a 45° cycling of the previous one. The maximum norm is then applied pixelwise to the eight resulting **compass gradients** to produce the image to be thresholded. The methodology is outlined in the block diagram of Fig. 3.10 (where "MAG0" means maximum) and a commonly used compass mask set is shown in Fig. 3.11. Clearly, if convolution is not implemented efficiently, then compass-gradient edge detection will be time consuming.

3.4 Linear Matched Filtering

Windowed convolution can be used to detect objects within an image. The idea is to convolve an input image with a mask that matches the gray-level configuration of the object to be detected. The result is a **matched filter** that gives back high values when the mask is located on top of the object of interest and lower values elsewhere.

The geometric intuition behind matched filtering can be seen in Eq. 3.2. Suppose g is a mask centered at the origin and it represents an object in the image centered at x. $f(x+t)$ represents the signal $f(t)$ shifted so that the object at x is now centered at the origin. Assuming the object and the mask are congruent, there is a match at the origin and this match will show up as a large value of $h(x)$. When noise is insignificant, the method works quite well.

To illustrate matched filtering in one dimension, consider the signal $f(t)$

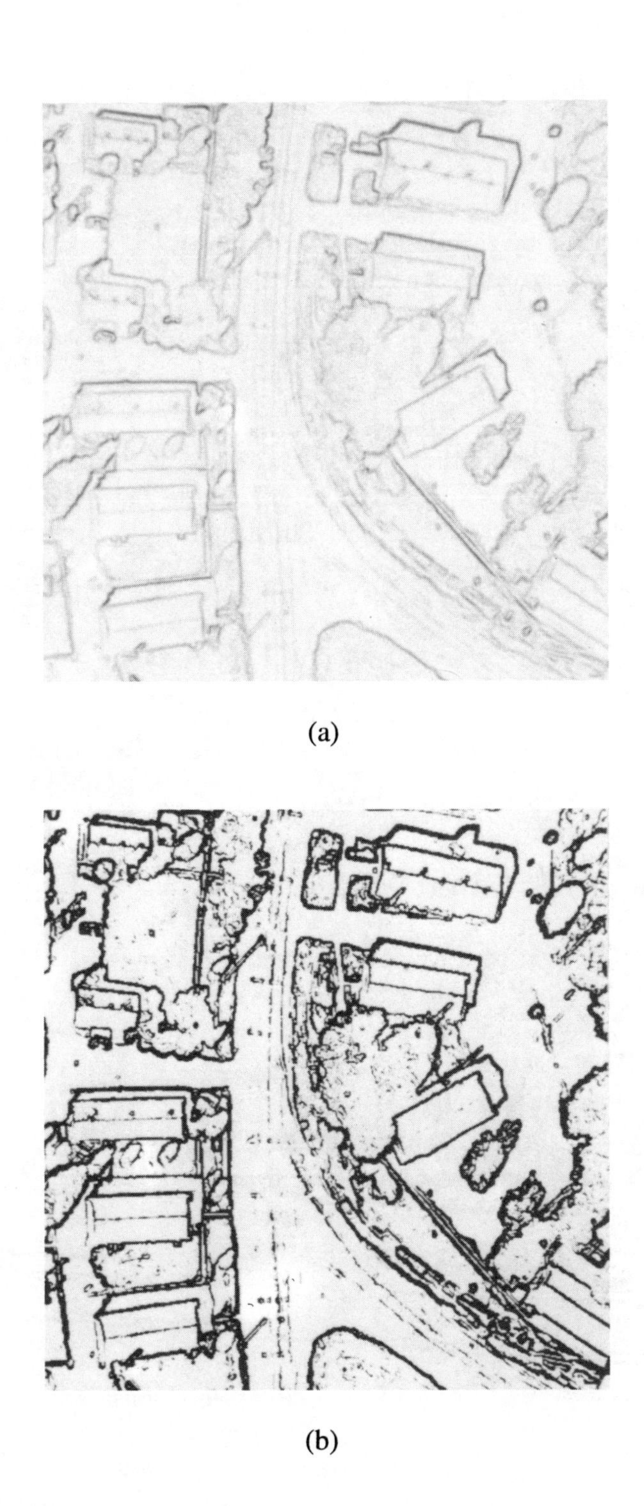

(a)

(b)

Figure 3.8: Aerial scene: (a) gradient image; (b) edge image.

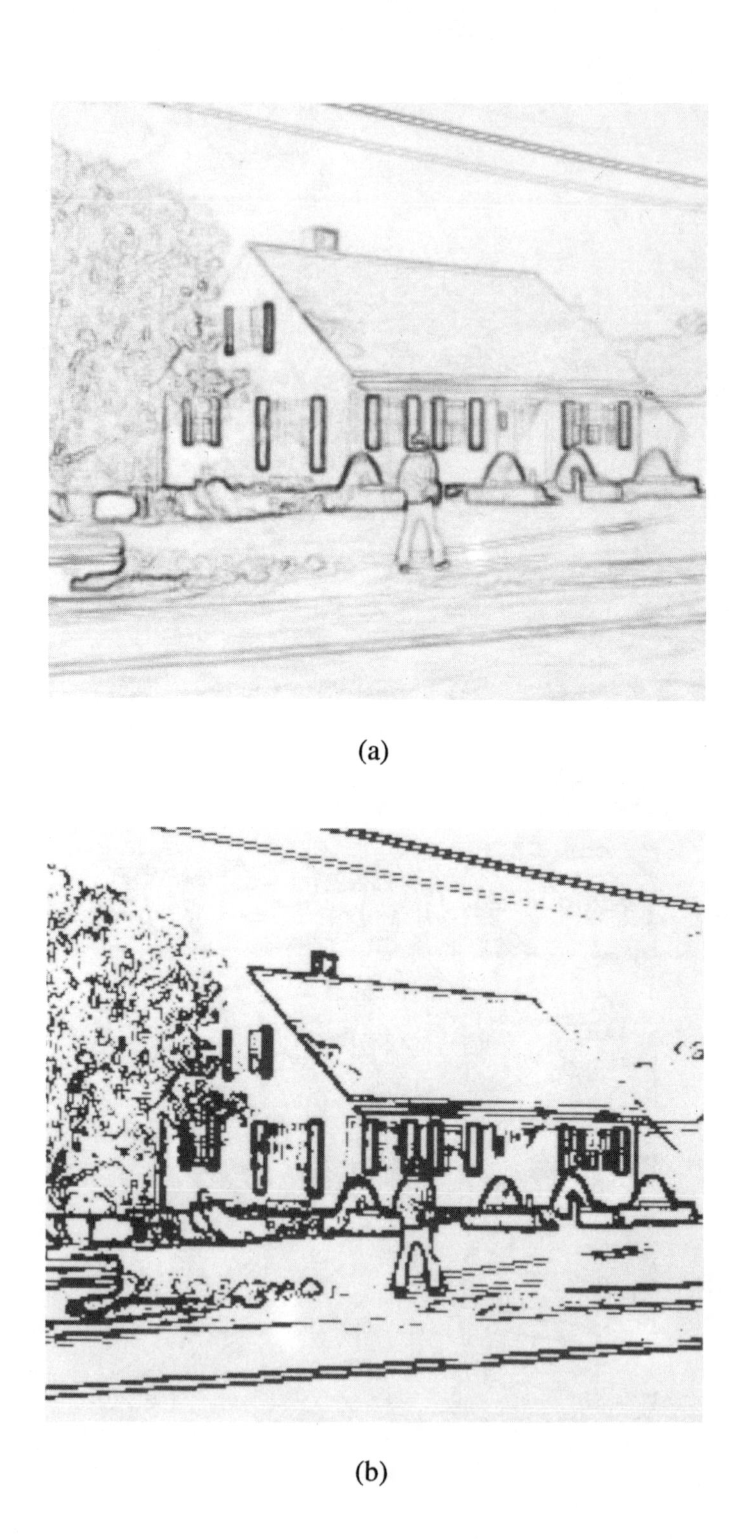

(a)

(b)

Figure 3.9: House scene: (a) gradient image; (b) edge image.

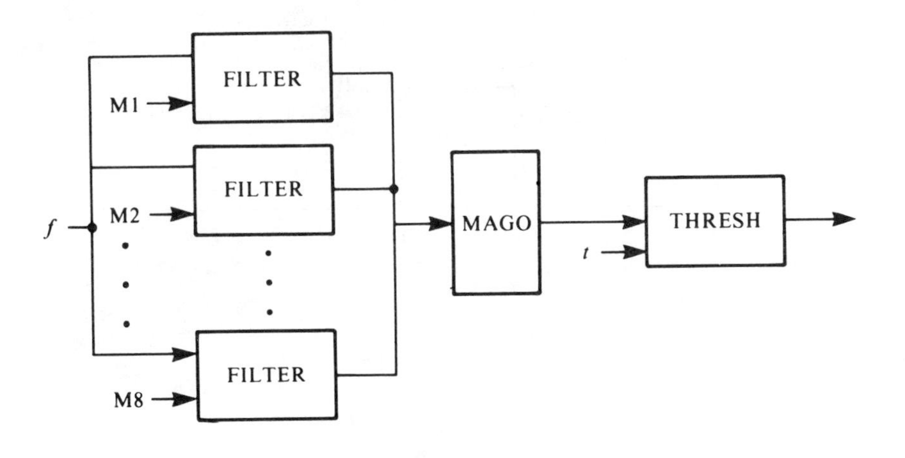

Figure 3.10: Compass-gradient edge detection.

$$
\begin{bmatrix} -1 & 0 & 1 \\ -1 & \mathbf{0} & 1 \\ -1 & 0 & 1 \end{bmatrix}
\begin{bmatrix} 0 & 1 & 1 \\ -1 & \mathbf{0} & 1 \\ -1 & -1 & 0 \end{bmatrix}
\begin{bmatrix} 1 & 1 & 1 \\ 0 & \mathbf{0} & 0 \\ -1 & -1 & -1 \end{bmatrix}
\begin{bmatrix} 1 & 1 & 0 \\ 1 & \mathbf{0} & -1 \\ 0 & -1 & -1 \end{bmatrix}
$$

$$
\begin{bmatrix} 1 & 0 & -1 \\ 1 & \mathbf{0} & -1 \\ 1 & 0 & -1 \end{bmatrix}
\begin{bmatrix} 0 & -1 & -1 \\ 1 & \mathbf{0} & -1 \\ 1 & 1 & 0 \end{bmatrix}
\begin{bmatrix} -1 & -1 & -1 \\ 0 & \mathbf{0} & 0 \\ 1 & 1 & 1 \end{bmatrix}
\begin{bmatrix} -1 & -1 & 0 \\ -1 & \mathbf{0} & 1 \\ 0 & 1 & 1 \end{bmatrix}
$$

Figure 3.11: Set of compass masks.

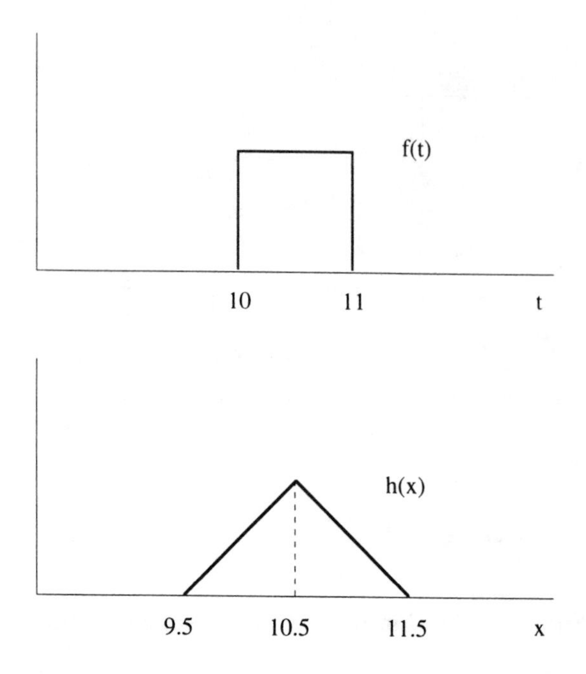

Figure 3.12: Matched filtering in one dimension.

shown in Fig. 3.12. Let the mask g be defined by $g(t) = 1$ for $-0.5 \leq t \leq 0.5$ and $g(t) = 0$ otherwise. Application of Eq. 3.2 yields

$$h(x) = \begin{cases} x - 9.5, & \text{if } 9.5 \leq x \leq 10.5 \\ 11.5 - x, & \text{if } 10.5 \leq x \leq 11.5 \\ 0, & \text{otherwise} \end{cases} \tag{3.10}$$

which, as shown in Fig. 3.12, is maximized at $x = 10.5$.

Figure 3.13 illustrates matched filtering. Part (a) shows a "y" template, (b) shows a binary (0 and 255 gray-level) text image, (c) shows convolution with the "y" template, and (d) shows the binary output of thresholding the convolution image. Note the locations of the "y"s in the original image.

3.5 Convolution Algorithm

To have a basis for analysis of implementation efficiency, we need to postulate some basic sequential computation scheme. This will provide us with a benchmark system to quantify execution times. We do not wish to go deeply

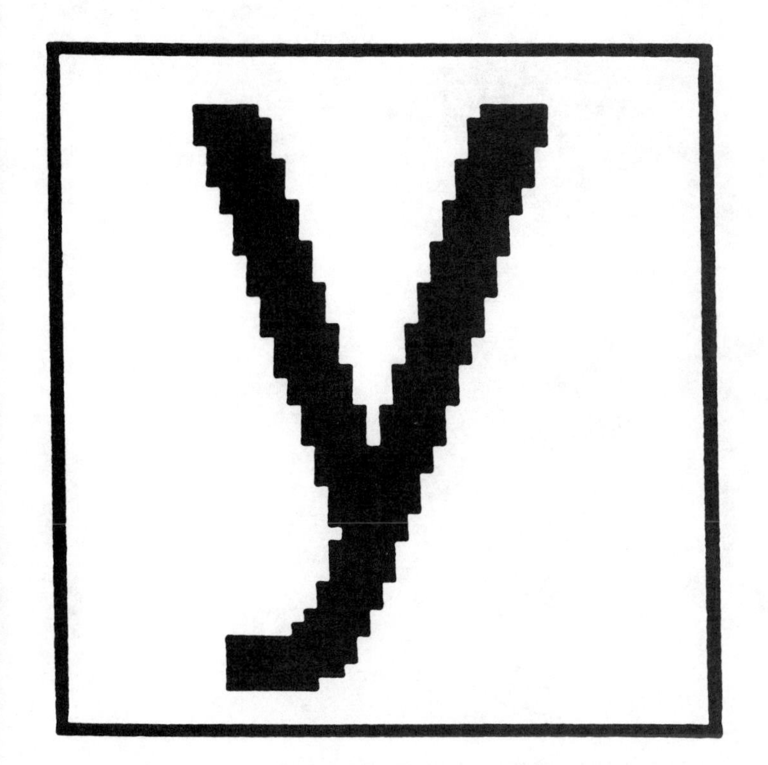

Figure 3.13: Matched filter detection: (a) template, (b) text image.

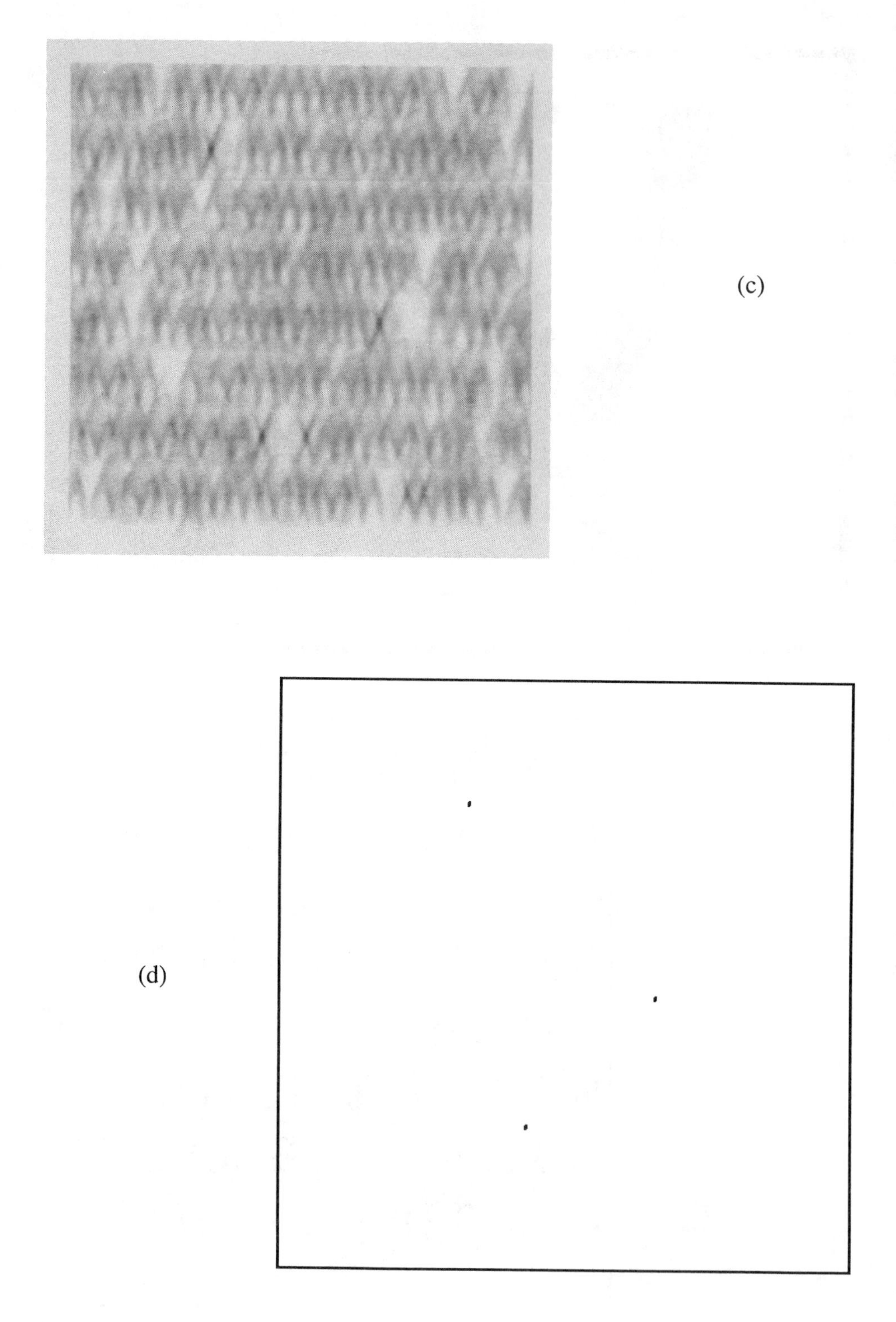

Figure 3.13: Matched filter detection: (c) convolution image, (d) thresholded convolution.

into computer organization. From the machine perspective, we shall remain at a rather high level, working at the assembly level with an instruction set tailored to the specific imaging algorithm under consideration. This means that we shall only consider macro-operations such as ADD and MOVE, not the micro–operations between registers that compose the macro-operations. Consequently, actual speeds of implementation will not be discussed because these are dependent on the micro-operation cycles that make up the macro-operations (as well as propagation delays, bus speeds, and other hardware related timings). Unless, otherwise stated, each macro-operation is assumed to take a single clock cycle, and we count these.

For our basic computer, there is an arithmetic-logic unit (ALU), two CPU registers, and sequential random-access memory. Later on we will consider hardware variations, especially parallel hardware, but for now the hardware is limited to a simple single-instruction-single-data (SISD) format and memory is not modular. Each instruction will involve an operation and two locations, either registers or memory addresses. For instance `[ADD; R1, R2]` means to add the contents of registers `R1` and `R2` and to store the result in register `R1`. Symbolically this instruction can be interpreted as `R1` $\longleftarrow$ `R1` + `R2`. One operand can be moved from memory in any given clock cycle. This allows us to add the operand from memory location `A` to the operand in register `R1`. `[ADD; R1, A]` means to add the contents of register `R1` and memory location `A` and to store the result in register `R1`. Symbolically, `R1` $\longleftarrow$ `R1` + `A`. The convolution algorithm will employ an instruction set composed of three macro-operations:

```
[MOVE; R1, A]:  R1 ⟵ A
[ADD; R1, R2]:  R1 ⟵ R1 + R2          (3.11)
[MULT; R1; A]:  R1 ⟵ R1 × A
```

As it is written, Eq. 3.1 provides a computer algorithm. One simply needs to apply the multiplications and additions as indicated. To avoid complicated notation, we employ the signal (one-dimensional) form of digital convolution to illustrate such an implementation. For signals, Eq. 3.1 becomes

$$h(i) = \sum_{r=-m}^{m} g(r)f(i+r). \tag{3.12}$$

We assume f is defined over the discrete points $0, 1, \ldots, n$ and that it has been extended m points on either side to accommodate the boundary problem created by a mask of length $2m + 1$.

Address	Operand	Address	Operand
0	$g(-m)$	$3m+2$	$f(1)$
1	$g(-m+1)$	$\vdots$	$\vdots$
$\vdots$	$\vdots$	$4m$	$f(m-1)$
$m-1$	$g(-1)$	$4m+1$	$f(m)$
m	$g(0)$	$4m+2$	$f(m+1)$
$m+1$	$g(1)$	$\vdots$	$\vdots$
$\vdots$	$\vdots$	$3m+n$	$f(n-1)$
$2m$	$g(m)$	$3m+n+1$	$f(n)$
$2m+1$	$f(-m)$	$3m+n+2$	$f(n+1)$
$2m+2$	$f(-m+1)$	$\vdots$	$\vdots$
$\vdots$	$\vdots$	$4m+n$	$f(n+m-1)$
$3m$	$f(-1)$	$4m+n+1$	$f(n+m)$
$3m+1$	$f(0)$		

Figure 3.14: Memory storage for digital signal convolution.

We assume $g(-m)$ through $g(m)$ are stored in memory addresses 0 through $2m$ and $f(-m)$ through $f(n+m)$ are stored in memory addresses $2m+1$ through $4m+n+1$. The output of the program consists of $h(0)$ through $h(n)$ and these values are to be stored in memory addresses k through $k+n$. The contents of memory addresses 0 through $4m+n+1$ are shown in Fig. 3.14.

A straightforward program based on Eq. 3.12 is shown in Fig. 3.15. Because the program was hand-generated, it omits code to increment and test the loop control variable. A compiler-generated version of this code, however, would include several instructions to increment and test a loop counter. We consider this aspect later in Chapter 8. To begin with, $g(-m)$ is moved from memory address 0 to `R1`. Then $g(-m)$ is multiplied by $f(-m)$ and the product is left in `R1`. Next, $g(-m+1)$ is moved to `R2`, $g(-m+1)$ is multiplied by $f(-m)$, and the product is left in `R2`. The contents of `R1` and `R2` are added and the sum is left in `R1`. Next, $g(-m+2)$ is moved to `R2` and $f(-m+2)$ is multiplied by $g(-m+2)$ with the product left in `R2`. The contents of `R1` and `R2` are then added with the sum being left in `R1`. The

```
BEGIN
MOVE   R1, 0            : compute h(0)
MULT   R1, 2m + 1
MOVE   R2, 1
MULT   R2, 2m + 2
ADD    R1, R2
MOVE   R2, 2
MULT   R2, 2m + 3
ADD    R1, R2
  .
  .
  .
MOVE   R2, 2m           : compute h(1)
MULT   R2, 4m + 1
ADD    R1, R2
MOVE   k, R1
MOVE   R1, 0
MULT   R1, 2m + 2
MOVE   R2, 1
MULT   R2, 2m + 3
ADD    R1, R2
MOVE   R2, 2
MULT   R2, 2m + 4
ADD    R1, R2
  .
  .
  .
MOVE   R2, 2m           : compute h(2)
MULT   R2, 4m + 2
ADD    R1, R2
MOVE   k + 1, R1
MOVE   R1, 0
MULT   R1, 2m + 3
  .
  .
  .
MOVE   R2, 2m           : compute h(n)
MULT   R2, 4m + n + 1
ADD    R1, R2
MOVE   k + n, R1
END
```

Figure 3.15: Convolution program.

process continues with R1 acting as the accumulator for the summation. After the sum $h(0)$ is completed in R1, it is moved to memory location k. The process then repeats for $h(1)$, $h(2), \ldots, h(n)$, with the convolution output being stored in memory locations k through $k + n$.

Each loop has $2m$ ADDs, $2m + 1$ MULTs, and $2m + 2$ MOVEs. Since there are $n + 1$ loops, there are $2m(n + 1)$ ADDs, $(2m + 1)(n + 1)$ MULTs, and $(2m + 2)(n + 1)$ MOVEs. Yet these numbers do not likely give the number of clock cycles. While the ALU can handle each ADD in one cycle and the control can move an operand in one cycle, unless there is a hardwired multiplier, each MULT probably involves a branching to a subroutine that will take some number of cycles before returning to the main program. Multiplications are computationally expensive and there is a large number of them in a direct implementation of convolution. Increasing computation efficiency for convolution will be addressed in subsequent chapters. A key real-time advantage of nonlinear processing is the absence of costly multiplication operations.

An important point needs to be made here. The above implementation, which is close to optimal in terms of real-time performance for an SISD architecture, was made directly in assembly language. If the algorithm was coded directly from Eq. 3.1 in a high-level language such as C, the code emitted by the compiler would be much slower. This arises from the fact that most compilers would not be able to calculate the absolute addresses of the data at compile time because of the use of the variable indices. Rather, the compiler will make use of register indirect, indirect, and even double indirect instructions. The unfortunate shortcomings of compiler-generated code, however, present opportunities for optimization techniques within the high-level language that will yield code that is nearly optimal. These techniques will be discussed in Chapter 8.

Chapter 4

Compression by Matrix Transforms

Digital images often contain very large amounts of information and a basic task of digital image processing is to compress the amount of data to be stored or transmitted by finding a suitable coding of the data. Once coded, the data can be more efficiently stored or transmitted and then decoded later for either display or processing. The present chapter introduces transform coding and illustrates the manner in which efficient algorithms can be used to reduce processing time for an image processing task employing matrix multiplication.

Often there is significant correlation between nearby pixels in a digital image and a basic approach to exploiting this correlation is application of an orthogonal matrix transform. Typically, an image is blocked, say into 8×8 or 16×16 square blocks, each block is transformed in such a manner that allows compression of the number of data entries, and, when desired, an inverse transform is applied to produce a recovered image. If there is no loss of information, then the coding is **lossless** and inversion yields the exact image that was compressed; if, as will be the case we study here, some (hopefully small) amount of information is lost upon compression, then the coding is **lossy** and the image recovered by inversion will be an estimate of the original. Lossy compression is useful if it results in a significant amount of compression and, upon decompression, the recovered image is suitable for whatever task is at hand. A block diagram for a transform-coding system is given in Fig. 4.1 Since our main purpose is to describe some computational aspects of matrix-transform compression, we leave the quantization and coding that occur prior

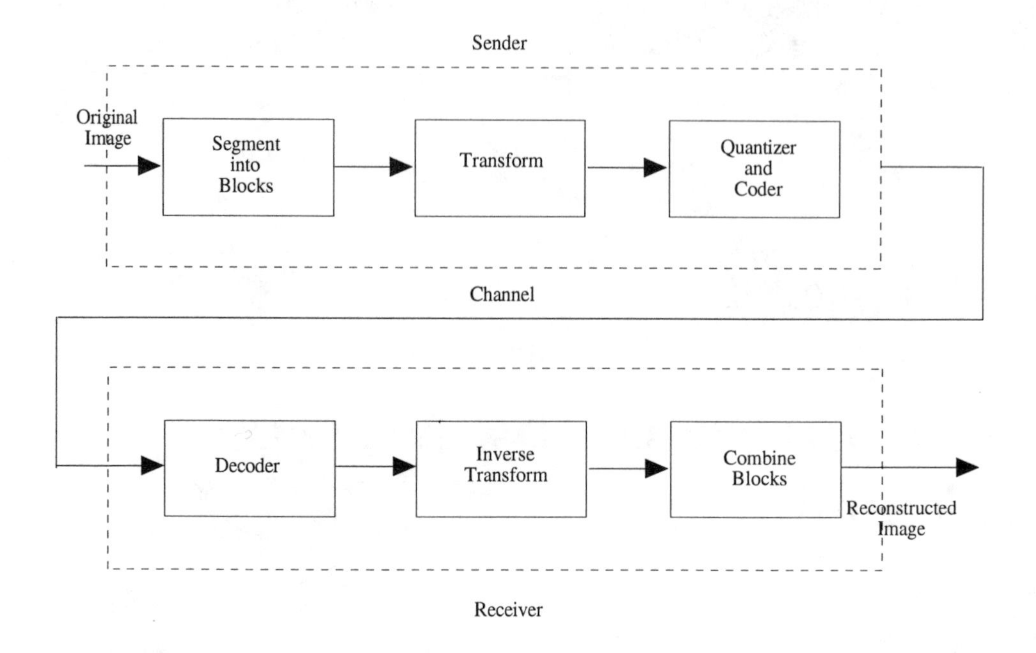

Figure 4.1: Block diagram for transform coding.

to the data entering a communication channel and the decoding subsequent to the data passing through the channel to a text on image compression.

4.1 Hadamard Transform

The rows of an orthogonal matrix form a basis set of orthogonal vectors and multiplying a vector by the matrix produces a vector whose entries correspond to a representation relative to this basis set. An orthogonal matrix is useful for compression if, after multiplication of a data vector by the matrix, components of the resulting vector are less correlated (less redundant) and total energy is concentrated in a small number of entries.

We explain the method via the **Hadamard transform**. For $q = 2^n, n = 1, 2, 3, \ldots$, the Hadamard matrix of size $N = 2^n$ can be defined recursively by

$$H(2q) = \frac{1}{\sqrt{2}} \begin{bmatrix} H(q) & H(q) \\ H(q) & -H(q) \end{bmatrix}, \tag{4.1}$$

where recursion is initiated by $H(1) = (1)$. Equation 4.1 leads to a set of matrices whose entries are 1 or -1. For the purposes of coding, the rows are usually reorganized so there is increasing oscillation between 1 and -1 as one traverses down the rows. Thus, for $n = 1, 2$, and 3, $H(2)$, $H(4)$, and $H(8)$ are taken to be

$$H(2) = \frac{1}{\sqrt{2}} \begin{bmatrix} 1 & 1 \\ 1 & -1 \end{bmatrix}$$

$$H(4) = \frac{1}{2} \begin{bmatrix} 1 & 1 & 1 & 1 \\ 1 & 1 & -1 & -1 \\ 1 & -1 & -1 & 1 \\ 1 & -1 & 1 & -1 \end{bmatrix} \tag{4.2}$$

$$H(8) = \frac{1}{\sqrt{8}} \begin{bmatrix} 1 & 1 & 1 & 1 & 1 & 1 & 1 & 1 \\ 1 & 1 & 1 & 1 & -1 & -1 & -1 & -1 \\ 1 & 1 & -1 & -1 & -1 & -1 & 1 & 1 \\ 1 & 1 & -1 & -1 & 1 & 1 & -1 & -1 \\ 1 & -1 & -1 & 1 & 1 & -1 & -1 & 1 \\ 1 & -1 & -1 & 1 & -1 & 1 & 1 & -1 \\ 1 & -1 & 1 & -1 & -1 & 1 & -1 & 1 \\ 1 & -1 & 1 & -1 & 1 & -1 & 1 & -1 \end{bmatrix}.$$

The basis signals corresponding to $H(8)$ are given in Fig. 4.2.

For real orthonormal basis matrices, the matrix inverse is equal to its transpose. Thus, the inverse of a Hadamard matrix H is the matrix itself: $H^{-1} = H^T = H$. If X is a discrete vector, then the transformed vector is HX. Since $H^{-1} = H$, inversion is performed by applying H a second time: $X = HHX$.

As an example, consider the vectors X and Y defined via their transposes

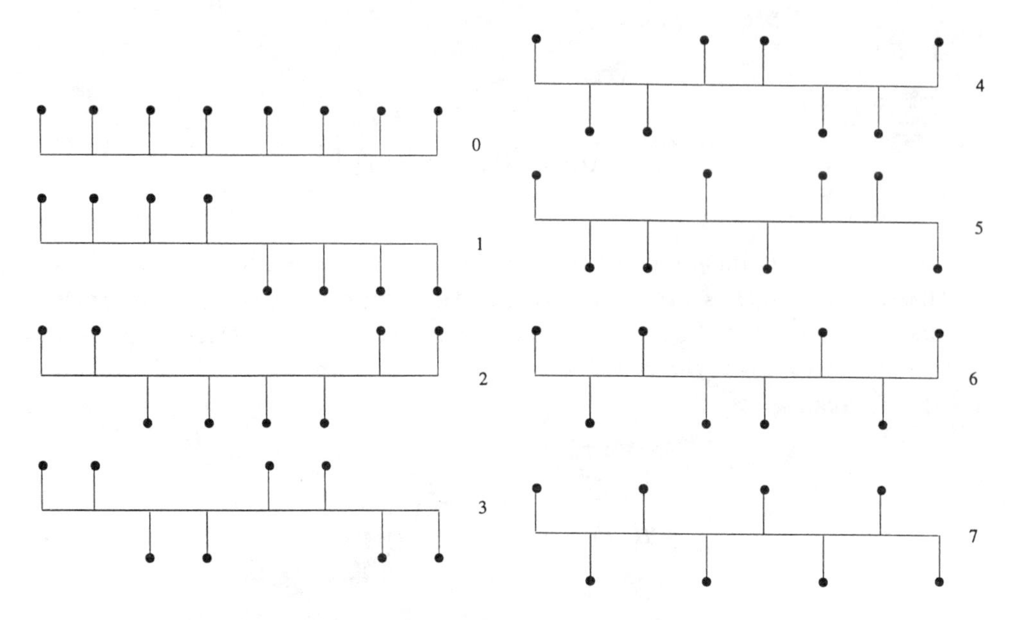

Figure 4.2: Basis signals for Hadamard transform with $N = 8$.

by $X^T = (8,7,9,6)$ and $Y^T = (12,10,6,4)$. Then

$$
\begin{aligned}
HX &= \tfrac{1}{2}\begin{bmatrix} 1 & 1 & 1 & 1 \\ 1 & 1 & -1 & -1 \\ 1 & -1 & -1 & 1 \\ 1 & -1 & 1 & -1 \end{bmatrix}\begin{bmatrix} 8 \\ 7 \\ 9 \\ 6 \end{bmatrix} = \begin{bmatrix} 15 \\ 0 \\ -1 \\ 2 \end{bmatrix} \\
HY &= \tfrac{1}{2}\begin{bmatrix} 1 & 1 & 1 & 1 \\ 1 & 1 & -1 & -1 \\ 1 & -1 & -1 & 1 \\ 1 & -1 & 1 & -1 \end{bmatrix}\begin{bmatrix} 12 \\ 10 \\ 6 \\ 4 \end{bmatrix} = \begin{bmatrix} 16 \\ 6 \\ 0 \\ 2 \end{bmatrix}.
\end{aligned} \qquad (4.3)
$$

X and Y can be retrieved from HX and HY by $X = H^{-1}HX = HHX$ and $Y = H^{-1}HY = HHY$. Compression is achieved by not transmitting (storing) X or Y but by transmitting only the lower components of HX and HY. For instance, if X_0 is obtained from HX by setting the last three components of HX equal to 0 (and the receiver knows this), then only the first component of X_0 needs to be transmitted. Decompression is achieved by multiplying X_0 by $H^{-1} = H$ to obtain an approximation of X, namely,

$$HX_0 = H \begin{bmatrix} 15 \\ 0 \\ 0 \\ 0 \end{bmatrix} = \begin{bmatrix} 7.5 \\ 7.5 \\ 7.5 \\ 7.5 \end{bmatrix}. \tag{4.4}$$

Compression followed by decompression has resulted in a constant vector whose entries are the average of the original vector's entries. If X is a relatively constant portion of a digital signal, then HX_0 is a constant estimate of that portion.

If the same compression-decompression scheme is used for Y, then

$$HY_0 = H \begin{bmatrix} 16 \\ 0 \\ 0 \\ 0 \end{bmatrix} = \begin{bmatrix} 8 \\ 8 \\ 8 \\ 8 \end{bmatrix}. \tag{4.5}$$

This time the decompression is not very good. It represents an averaging; however, the original vector Y is decreasing markedly.

Suppose rather than only taking the first component of HY_0, we take the first two components as the compressed vector. Denoting this compressed vector by Y_1, decompression yields

$$HY_1 = H \begin{bmatrix} 16 \\ 6 \\ 0 \\ 0 \end{bmatrix} = \begin{bmatrix} 11 \\ 11 \\ 5 \\ 5 \end{bmatrix}. \tag{4.6}$$

If the gap between the second and third components of Y represents a jump between fairly flat regions, then HY_1 provides a decent estimate of Y. Typically, a signal is blocked and the transform is applied to each block. Parts (a), (b), and (c) of Fig 4.3 depict the original two-block, eight-point signal (X, Y), the decompressed signal (HX_0, HY_0), and the decompressed signal (HX_0, HY_1). In the figure it is assumed that each discrete time point is at the center of a pixel and the line drawings depict the analog signals corresponding to the digital signals.

The preceding example illustrates a problem with lossy transform coding: how does one choose the compressed transform vector? In the **zonal method**, one simply fixes the set of components to be kept; in the **threshold method**, those components with values above some threshold are kept.

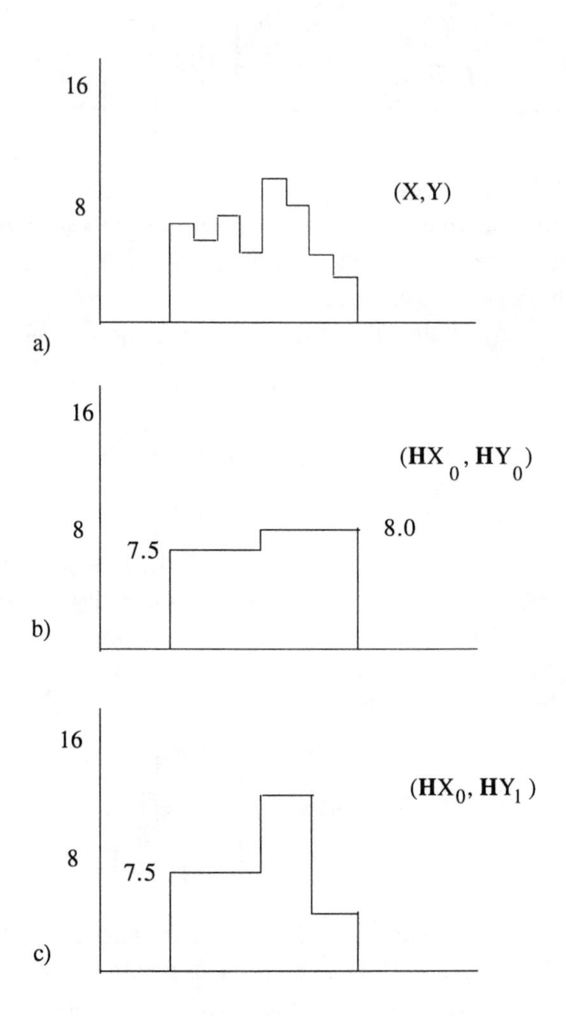

Figure 4.3: Compression and decompression using Hadamard transform.

The threshold method requires more overhead because the receiver must be informed as to which components are being kept. Ipso facto, overhead decreases the efficiency of the coding scheme.

In terms of linear systems, the rows of the Hadamard matrix form an orthonormal system and the transformed vector HX gives the coefficients of X relative to this system. Consequently, if $(HX)^T = (y_1, y_2, y_3, y_4)$, then

$$X = \frac{1}{2}\left(y_1 \begin{bmatrix} 1 \\ 1 \\ 1 \\ 1 \end{bmatrix} + y_2 \begin{bmatrix} 1 \\ 1 \\ -1 \\ -1 \end{bmatrix} + y_3 \begin{bmatrix} 1 \\ -1 \\ -1 \\ 1 \end{bmatrix} + y_4 \begin{bmatrix} 1 \\ -1 \\ 1 \\ -1 \end{bmatrix} \right). \tag{4.7}$$

The first component of HX contains constant signal content and succeeding coefficients contain increasing oscillatory content, with the last representing the highest oscillatory content. This is analogous to frequency representation with Fourier series. The Hadamard coefficients are called **sequency coefficients**. Compression by eliminating higher-order coefficients of HX amounts to suppression of fine detail and preservation of less varying signal content. The specific detail loss is a function of the orthonormal vectors in the Hadamard system.

Adaptation of Hadamard compression to images is accomplished by blocking the image into 2^n by 2^n blocks and applying the transform to each block. If B is such a block and H is the Hadamard matrix of dimension 2^n, then the transformed block is

$$B' = HBH^T = HBH. \tag{4.8}$$

Inversion is accomplished by

$$B = H^T B' H = HB'H. \tag{4.9}$$

In one dimension, energy is packed into the lower-order vector components; in two dimensions, it is packed into the upper left corner of B'. For images, compression is accomplished by keeping only some set of upper left components. The transform provides a representation in terms of orthonormal basis images. These are graphically depicted in Fig. 4.4 for the 4×4 case, where white and black denote $+1$ and -1, respectively.

Relative to real-time processing, an advantage of the Hadamard transform over other transforms is that multiplication by the Hadamard matrix can be done without arithmetic multiplications; it requires only additions

Figure 4.4: Two-dimensional basis images for Hadamard transform.

and subtractions. The absence of multiplications has significant benefits for digital computing. Nonetheless, it is advantageous to further reduce the amount of computation since both compression and decompression require a large number of matrix multiplications.

A **fast Hadamard transform** is achieved via factorization. Let

$$A = \begin{bmatrix} 1 & 1 & 0 & 0 & 0 & 0 & 0 & 0 \\ 0 & 0 & 1 & 1 & 0 & 0 & 0 & 0 \\ 0 & 0 & 0 & 0 & 1 & 1 & 0 & 0 \\ 0 & 0 & 0 & 0 & 0 & 0 & 1 & 1 \\ 1 & -1 & 0 & 0 & 0 & 0 & 0 & 0 \\ 0 & 0 & 1 & -1 & 0 & 0 & 0 & 0 \\ 0 & 0 & 0 & 0 & 1 & -1 & 0 & 0 \\ 0 & 0 & 0 & 0 & 0 & 0 & 1 & -1 \end{bmatrix}. \tag{4.10}$$

Then, for 8 dimensions, the Hadamard matrix has the factorization

$$H(8) = A \times A \times A. \tag{4.11}$$

Applying $H(8)$ directly requires 49 additions or subtractions; applying A requires 8 additions or subtractions. Since A must be applied 3 times, factorization results in a total of 24 additions or subtractions. Generally, for the $N \times N$ Hadamard matrix, using the factorization requires $N\log_2 N$ additions and subtractions.

4.2 Discrete Fourier Transform

A second transform used in coding schemes is the discrete Fourier transform (DFT). It is a discrete version of the continuous Fourier transform and therefore provides a frequency-based spectral representation of the input signal or image. Since highly correlated data have little energy at high frequencies, the DFT packs information into lower-order components. As one might expect from the frequency interpretation, compression suppresses high-frequency content in favor of low frequencies. The complex nature of the DFT causes both storage and computation problems.

The (k, m) entry of the $N \times N$ matrix for the **discrete Fourier transform** as applied to discrete signals is given by

$$c_{km} = N^{\frac{-1}{2}} e^{\frac{-2\pi i k m}{N}} = N^{\frac{-1}{2}} \left[\cos\left(\frac{2\pi k m}{N}\right) - i \sin\left(\frac{2\pi k m}{N}\right)\right] \tag{4.12}$$

for $k, m = 0, 1, \ldots, N-1$, where $i = \sqrt{-1}$. Figure 4.5 gives the real and imaginary DFT matrices for $N = 8$ and Fig. 4.6 shows the corresponding basis signals. Letting F denote the DFT transform matrix defined by Eq. 4.12, the inverse matrix is

$$F^{-1} = [F^*]^T = F^*, \tag{4.13}$$

where the asterisk denotes the complex conjugate. Note that $F^T = F$ and $[F^*]^T = F^*$.

For images, blocking is done as in the case of the Hadamard transform. For a block B, the forward and inversion transforms are given by

$$B' = FBF^T = FBF, \tag{4.14}$$

$$B = [F^*]^T B' F^* = F^* B' F^*, \tag{4.15}$$

respectively. Compression is achieved by only taking the lower-order spatial frequencies in the upper left corner of B'.

Owing to the large number of multiplications to be performed, direct application of the DFT matrix is computationally expensive and much research has gone into efficient implementation. Here we describe the basic **fast Fourier transform (FFT)** factorization for $N = 8$.

$$
Re(F) = \begin{bmatrix}
0.3536 & 0.3536 & 0.3536 & 0.3536 & 0.3536 & 0.3536 & 0.3536 & 0.3536 \\
0.3536 & 0.2500 & 0.0000 & -0.2500 & -0.3536 & -0.2500 & 0.0000 & 0.2500 \\
0.3536 & 0.0000 & -0.3536 & 0.0000 & 0.3536 & 0.0000 & -0.3536 & 0.0000 \\
0.3536 & -0.2500 & 0.0000 & 0.2500 & -0.3536 & 0.2500 & 0.0000 & -0.2500 \\
0.3536 & -0.3536 & 0.3536 & -0.3536 & 0.3536 & -0.3536 & 0.3536 & -0.3536 \\
0.3536 & -0.2500 & 0.0000 & 0.2500 & -0.3536 & 0.2500 & 0.0000 & -0.2500 \\
0.3536 & 0.0000 & -0.3536 & 0.0000 & 0.3536 & 0.0000 & -0.3536 & 0.0000 \\
0.3536 & 0.2500 & 0.0000 & -0.2500 & -0.3536 & -0.2500 & 0.0000 & 0.2500
\end{bmatrix}
$$

$$
Im(F) = \begin{bmatrix}
0.0000 & 0.0000 & 0.0000 & 0.0000 & 0.0000 & 0.0000 & 0.0000 & 0.0000 \\
0.0000 & 0.2500 & 0.3536 & 0.2500 & 0.0000 & -0.2500 & -0.3536 & -0.2500 \\
0.0000 & 0.3536 & 0.0000 & -0.3536 & 0.0000 & 0.3536 & 0.0000 & -0.3536 \\
0.0000 & 0.2500 & -0.3536 & 0.2500 & 0.0000 & -0.2500 & 0.3536 & -0.2500 \\
0.0000 & 0.0000 & 0.0000 & 0.0000 & 0.0000 & 0.0000 & 0.0000 & 0.0000 \\
0.0000 & -0.2500 & 0.3536 & -0.2500 & 0.0000 & 0.2500 & -0.3536 & 0.2500 \\
0.0000 & -0.3536 & 0.0000 & 0.3536 & 0.0000 & -0.3536 & 0.0000 & 0.3536 \\
0.0000 & -0.2500 & -0.3536 & -0.2500 & 0.0000 & 0.2500 & 0.3536 & 0.2500
\end{bmatrix}
$$

Figure 4.5: Real and imaginary parts of DFT matrix, $N = 8$.

Let $W = e^{-2\pi i/N}$. Ignoring the $N^{1/2}$ scaling factor, which need not be applied in coding and is not important for computational efficiency, the DFT matrix possesses the factorization

$$F = A_3 \times A_2 \times A_1, \tag{4.16}$$

where A_1, A_2, and A_3 are given in Fig. 4.7. Given an input vector X, this decomposition results in the computation

$$X \longrightarrow A_1X \longrightarrow A_2A_1X \longrightarrow A_3A_2A_1X, \tag{4.17}$$

which involves sparse matrices and, in comparison to the original DFT matrix, few multiplications.

Representation via the DFT is closely related to representation via Fourier series. To give a visually perceptible illustration of how increasing compression results in decreasing accuracy, we provide an example based on Fourier series representation of a two–dimensional line drawing defined via the chain code (Fig. 4.8). The line drawing of Fig. 4.9 results from the chain code representation 0005676644422123. The line drawing itself is represented by a parameterization $x = x(t)$ and $y = y(t)$. Both $x(t)$ and $y(t)$ can be expressed

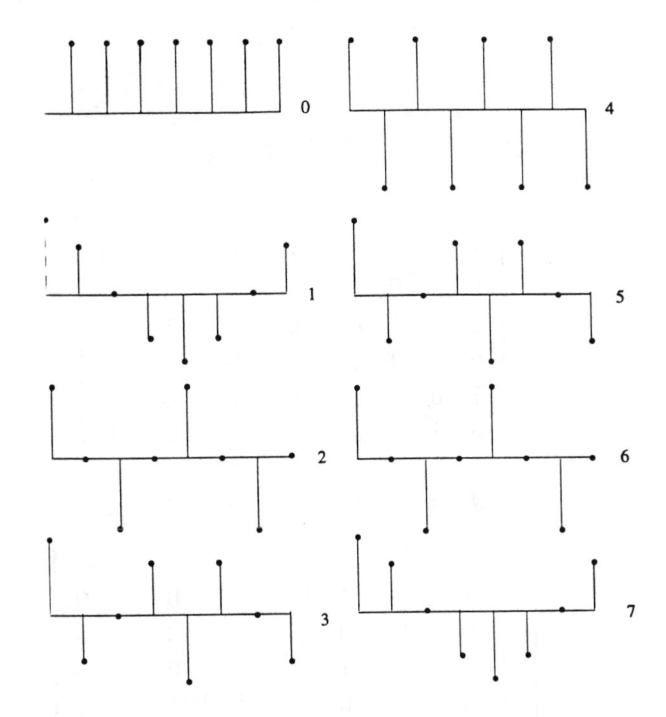

Figure 4.6: Basis signals for DFT with N = 8: (a) Real part.

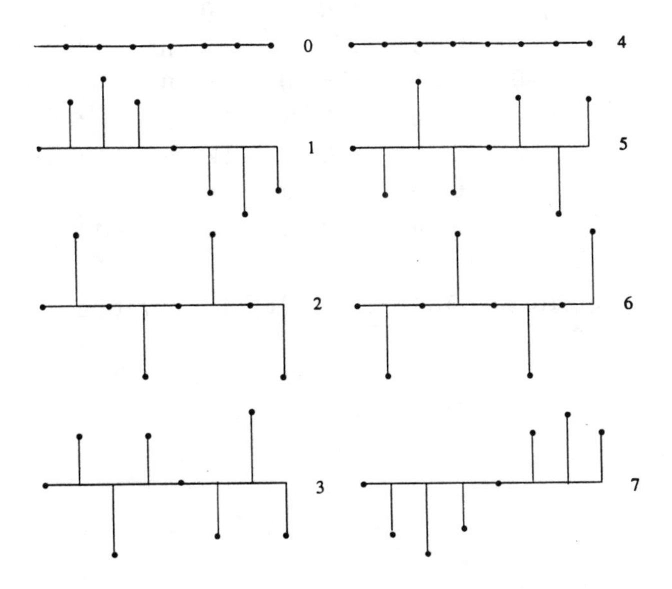

Figure 4.6: Basis signals for DFT with N = 8: (b) Imaginary part.

$$A_1 = \begin{bmatrix} 1 & 0 & 0 & 0 & 1 & 0 & 0 & 0 \\ 0 & 1 & 0 & 0 & 0 & 1 & 0 & 0 \\ 0 & 0 & 1 & 0 & 0 & 0 & 1 & 0 \\ 0 & 0 & 0 & 1 & 0 & 0 & 0 & 1 \\ 1 & 0 & 0 & 0 & -1 & 0 & 0 & 0 \\ 0 & 1 & 0 & 0 & 0 & -1 & 0 & 0 \\ 0 & 0 & 1 & 0 & 0 & 0 & -1 & 0 \\ 0 & 0 & 0 & 1 & 0 & 0 & 0 & -1 \end{bmatrix}$$

$$A_2 = \begin{bmatrix} 1 & 0 & 1 & 0 & 0 & 0 & 0 & 0 \\ 0 & 1 & 0 & 1 & 0 & 0 & 0 & 0 \\ 1 & 0 & -1 & 0 & 0 & 0 & 0 & 0 \\ 0 & 1 & 0 & -1 & 0 & 0 & 0 & 0 \\ 0 & 0 & 0 & 0 & 1 & 0 & W^2 & 0 \\ 0 & 0 & 0 & 0 & 0 & 1 & 0 & W^2 \\ 0 & 0 & 0 & 0 & 1 & 0 & -W^2 & 0 \\ 0 & 0 & 0 & 0 & 0 & 1 & 0 & -W^2 \end{bmatrix}$$

$$A_3 = \begin{bmatrix} 1 & 1 & 0 & 0 & 0 & 0 & 0 & 0 \\ 1 & -1 & 0 & 0 & 0 & 0 & 0 & 0 \\ 0 & 0 & 1 & W^2 & 0 & 0 & 0 & 0 \\ 0 & 0 & 1 & -W^2 & 0 & 0 & 0 & 0 \\ 0 & 0 & 0 & 0 & 1 & W & 0 & 0 \\ 0 & 0 & 0 & 0 & 1 & -W & 0 & 0 \\ 0 & 0 & 0 & 0 & 0 & 0 & 1 & W^3 \\ 0 & 0 & 0 & 0 & 0 & 0 & 1 & -W^3 \end{bmatrix}$$

Figure 4.7: Matrices for DFT factorization.

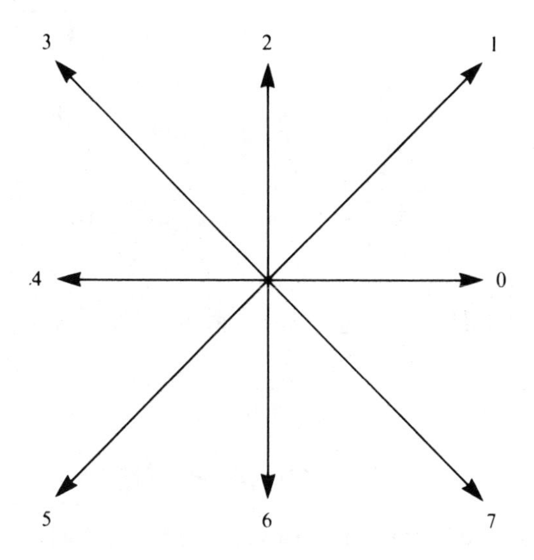

Figure 4.8: Chain code.

as Fourier series, for instance,

$$x(t) = a_0 + \sum_{n=1}^{\infty} \left(a_n \cos \frac{2n\pi t}{T} + b_n \sin \frac{2n\pi t}{T} \right). \tag{4.18}$$

If $X_N(t)$ and $Y_N(t)$ are the truncations to N harmonics of the $x(t)$ and $y(t)$ Fourier series, respectively, then $X_N(t)$ and $Y_N(t)$ represent approximations of $x(t)$ and $y(t)$, the $N + 1$ Fourier coefficients for $X_N(t)$ and $Y_N(t)$ are compressed versions of the full Fourier coefficient sequences, and the curve $(X_N(t), Y_N(t))$ is an approximation of the original curve $(x(t), y(t))$. Figure 4.9 shows the first four harmonic approximations.

Let us define the x and y approximation errors by

$$e_x[N] = \max_t \mid x(t) - X_N(t) \mid \tag{4.19}$$

$$e_y[N] = \max_t \mid y(t) - Y_N(t) \mid \tag{4.20}$$

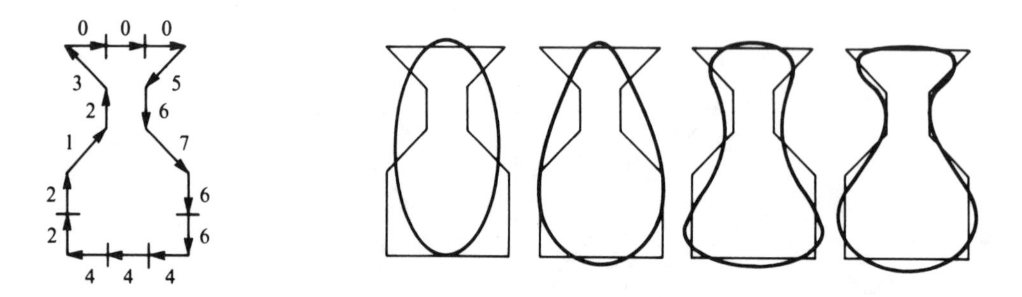

Figure 4.9: Line drawing and first four harmonic approximations.

and the overall error of approximation by

$$e[N] = \max\{e_x[N], e_y[N]\}. \tag{4.21}$$

As $N \to \infty$ (so the full Fourier series are used), $e[N] \to 0$. Figure 4.10 shows a chain-code line drawing, truncated Fourier series approximations for increasing N, and the error curve for increasing N.

4.3 Discrete Cosine Transform

While the Hadamard transform has natural computational advantages and can be employed for certain classes of images and the DFT represents a discrete version of a complete frequency-based orthonormal system, the transform of choice is usually the discrete cosine transform (DCT). It is close to statistically optimal for commonplace images possessing strong pixel-to-pixel correlation. Since optimal transforms are image dependent, the fact that the DCT is often close to optimal and not image dependent makes it an excellent candidate for many applications. Like the DFT, the DCT represents a discrete version of a frequency-based continuous transform, the DCT corresponding to the cosine transform. The basis functions for the DCT are digital samplings of the cosine basis functions. Like the fast DFT, there exists a fast DCT. Application to compression is similar to both the Hadamard transform and DFT.

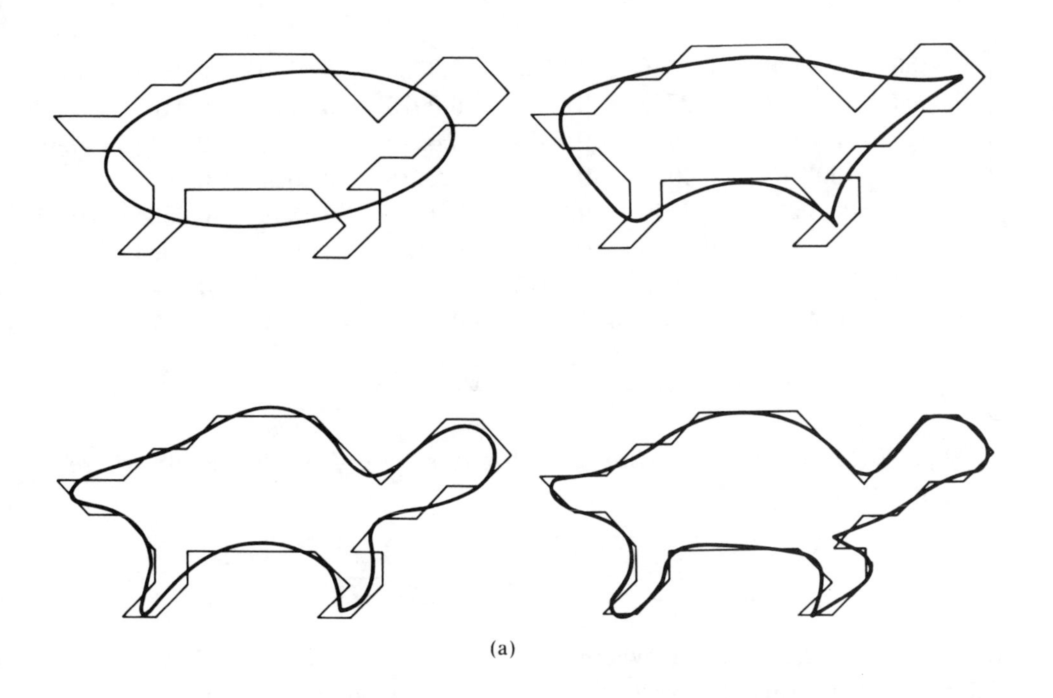

(a)

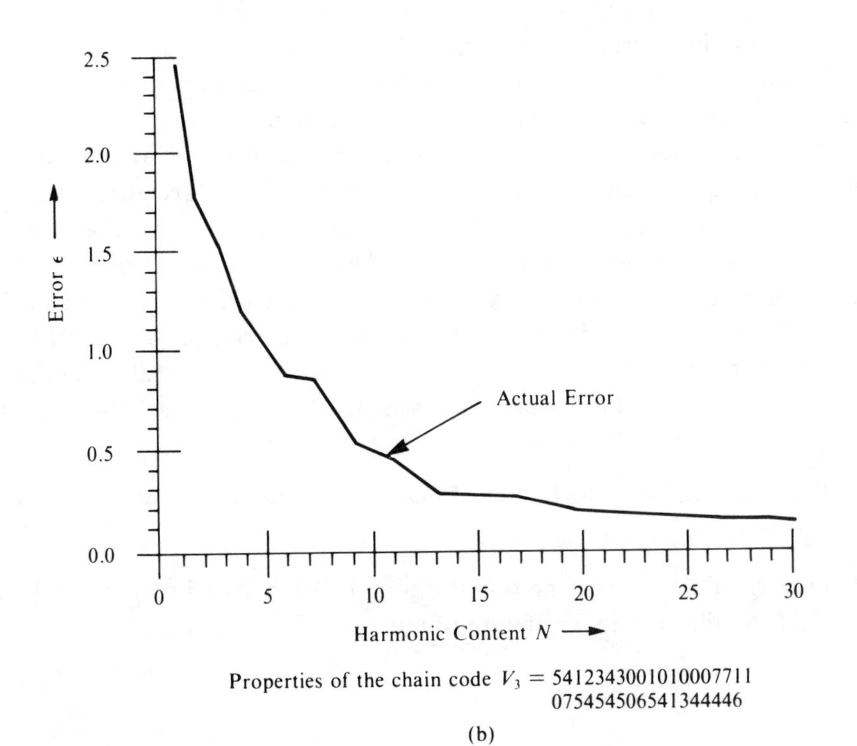

Properties of the chain code V_3 = 5412343001010007711
075454506541344446

(b)

Figure 4.10: Approximations and error for chain-coded raccoon.

$$C =
\begin{bmatrix}
0.3536 & 0.3536 & 0.3536 & 0.3536 & 0.3536 & 0.3536 & 0.3536 & 0.3536 \\
0.4904 & 0.4157 & 0.2778 & 0.0975 & -0.0975 & -0.2778 & -0.4157 & -0.4904 \\
0.4619 & 0.1913 & -0.1913 & -0.4619 & -0.4619 & -0.1913 & 0.1913 & 0.4619 \\
0.4157 & -0.0975 & -0.4904 & -0.2778 & 0.2778 & 0.4904 & 0.0975 & -0.4157 \\
0.3536 & -0.3536 & -0.3536 & 0.3536 & 0.3536 & -0.3536 & -0.3536 & 0.3536 \\
0.2778 & -0.4904 & 0.0975 & 0.4157 & -0.4157 & -0.0975 & 0.4904 & -0.2778 \\
0.1913 & -0.4619 & 0.4619 & -0.1913 & -0.1913 & 0.4619 & -0.4619 & 0.1913 \\
0.0975 & -0.2778 & 0.4157 & -0.4904 & 0.4904 & -0.4157 & 0.2778 & -0.0975
\end{bmatrix}$$

Figure 4.11: DCT matrix for $N = 8$.

The $N \times N$ **discrete cosine transform** is defined by the matrix whose components are given by

$$c_{km} = c_0 \left(\frac{2}{N}\right)^{\frac{1}{2}} \cos\left[\frac{m(k+\frac{1}{2})\pi}{N}\right] \tag{4.22}$$

for $k, m = 0, 1, 2, \ldots, N-1$, where $c_0 = 2^{-1/2}$ if $m = 0$ and $c_0 = 1$ if $m \neq 0$. There are two forms of the DCT, odd and even, having to do with the way a signal block is periodically extended. Equation 4.22 gives the even form of the DCT, which, for reasons having to do with block boundaries, generally provides superior performance. The DCT matrix for $N = 8$ is given in Fig. 4.11 and the corresponding basis signals are shown in Fig. 4.12. The two–dimensional basis images for $N = 8$ are shown in Fig. 4.13.

Often the DCT is employed in the JPEG baseline system, which, among other items, specifies blocking, coefficient normalization, quantization, and Huffman coding of the signal. **JPEG (Joint Photographic Experts Group)** is a lossy compression technique that is an industry standard for image information storage and retrieval. The overall conversion time for a typical square block of pixel data is constant; however, due to transmission of the Huffman-encoded bits, the ultimate size of the compressed pixel block, and thus its transmission time, varies, i.e., is not predictable (although it can be bounded by assuming a worst-case scenario). In any case, the following is a summary of the scheme:

1. Break the image into blocks of $k \times k$ pixels with positive integer gray values.

2. Perform a discrete cosine transform on this block and store the integral DCT coefficients in a corresponding $k \times k$ block array.

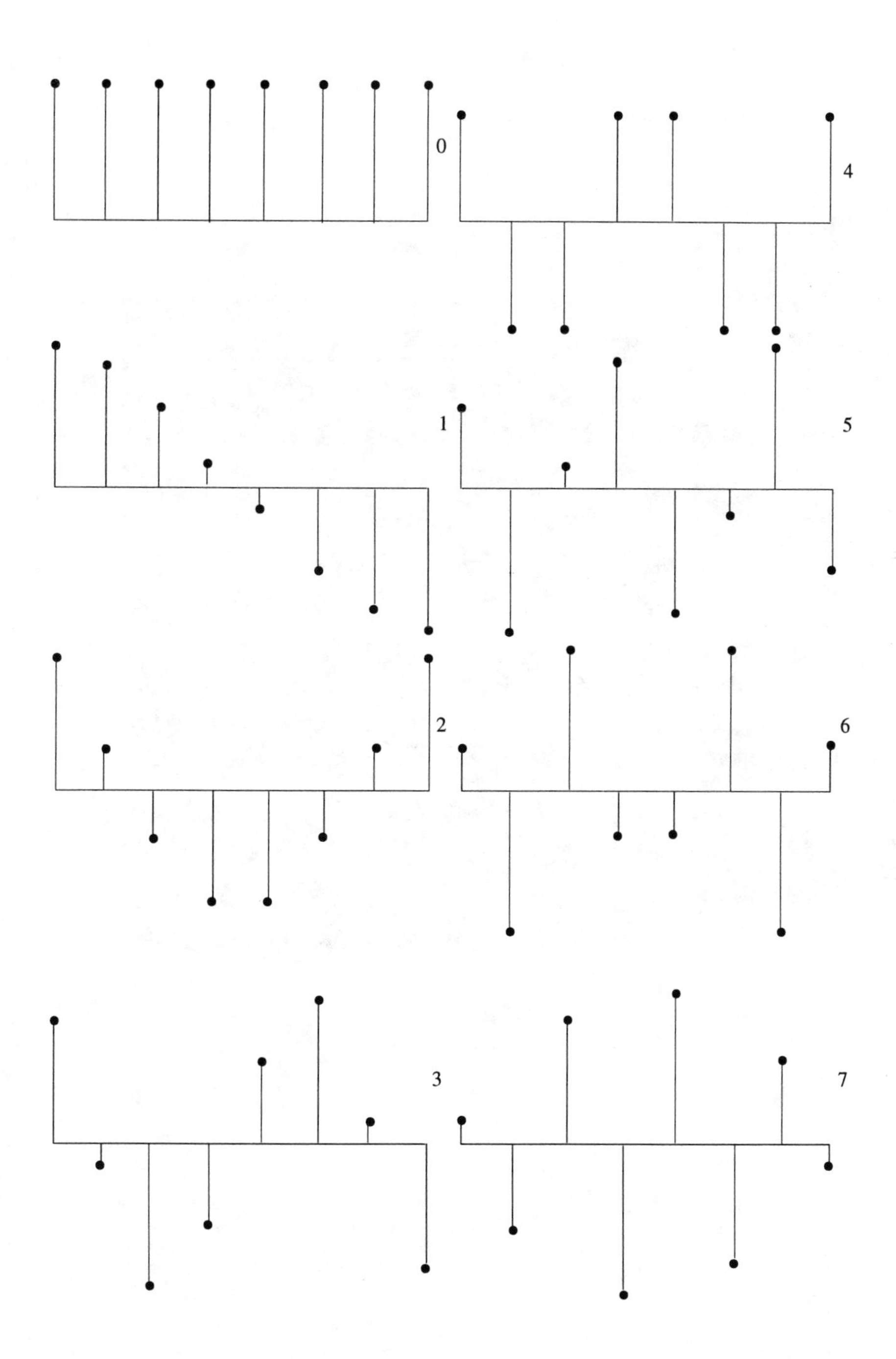

Figure 4.12: DCT basis signals for $N = 8$.

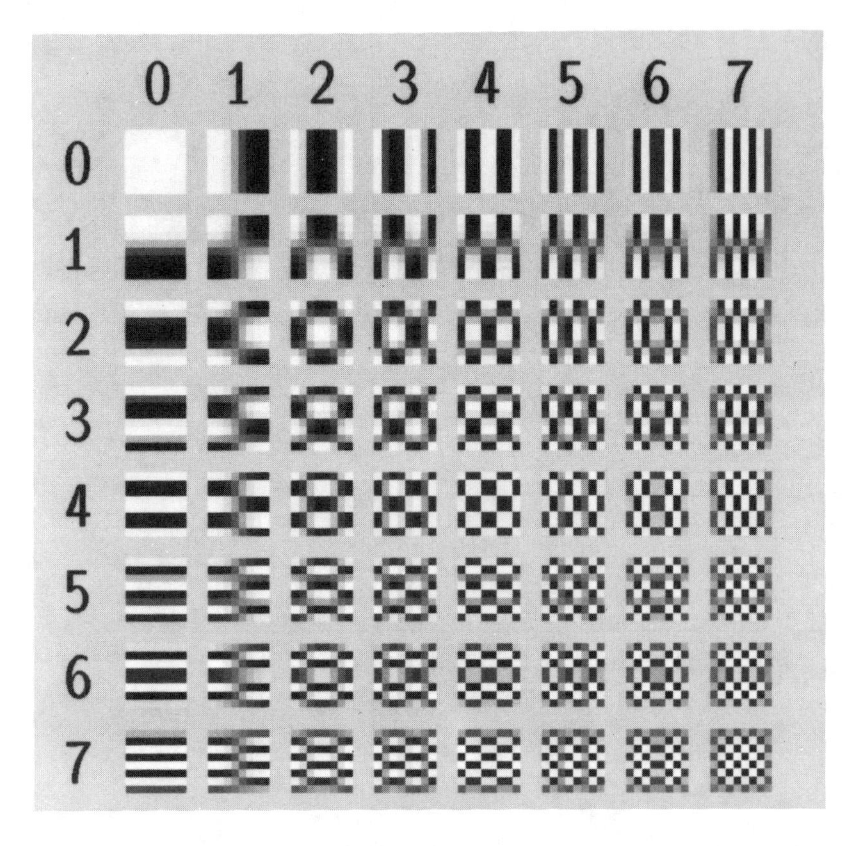

Figure 4.13: DCT basis images for $N = 8$.

3. Using a zig-zag run encoder, convert the $k \times k$ array of DCT coefficients into a column vector of length k^2 (zig-zag goes from left to right and top to bottom).

4. Quantize the column vector of DCT coefficients so that each DCT coefficient is converted to a quantized value located in a table of Huffman encoded values.

5. Huffman encode the quantized DCT coefficients.

6. Transmit encoding stream.

Figure 4.14 shows an original 512 × 512, 8-bit/pixel image and a 0.5-bit/pixel compressed image resulting from the JPEG DCT. Figure 4.15 shows blow-ups of the eye for the two images of Fig. 4.14.

The **MPEG (Motion Picture Engineers' Expert Group)** compression, used for motion pictures, essentially utilizes JPEG compression along with interframe compression, which discards pixel information that has not changed from the previous frame.

DCT encoding is deterministic (i.e., the number of floating and integer operations does not depend on the data) and there are numerous hardware chips that can perform this operation "quickly." Hence, it is well suited for real-time applications. We will subsequently discuss methods for improving DCT performance.

4.4 Quadtree Compression

Quadtree compression is an *ad hoc*, nonpredictive technique. In quadtrees, the image is divided recursively into four parts, stopping when each part is a constant value. If an image has a large number of regions with the same color, the compression ratio can be quite good. An illustration of quadtree representation for images is shown in Fig. 4.16.

Figure 4.14: Original and JPEG DCT compressed images.

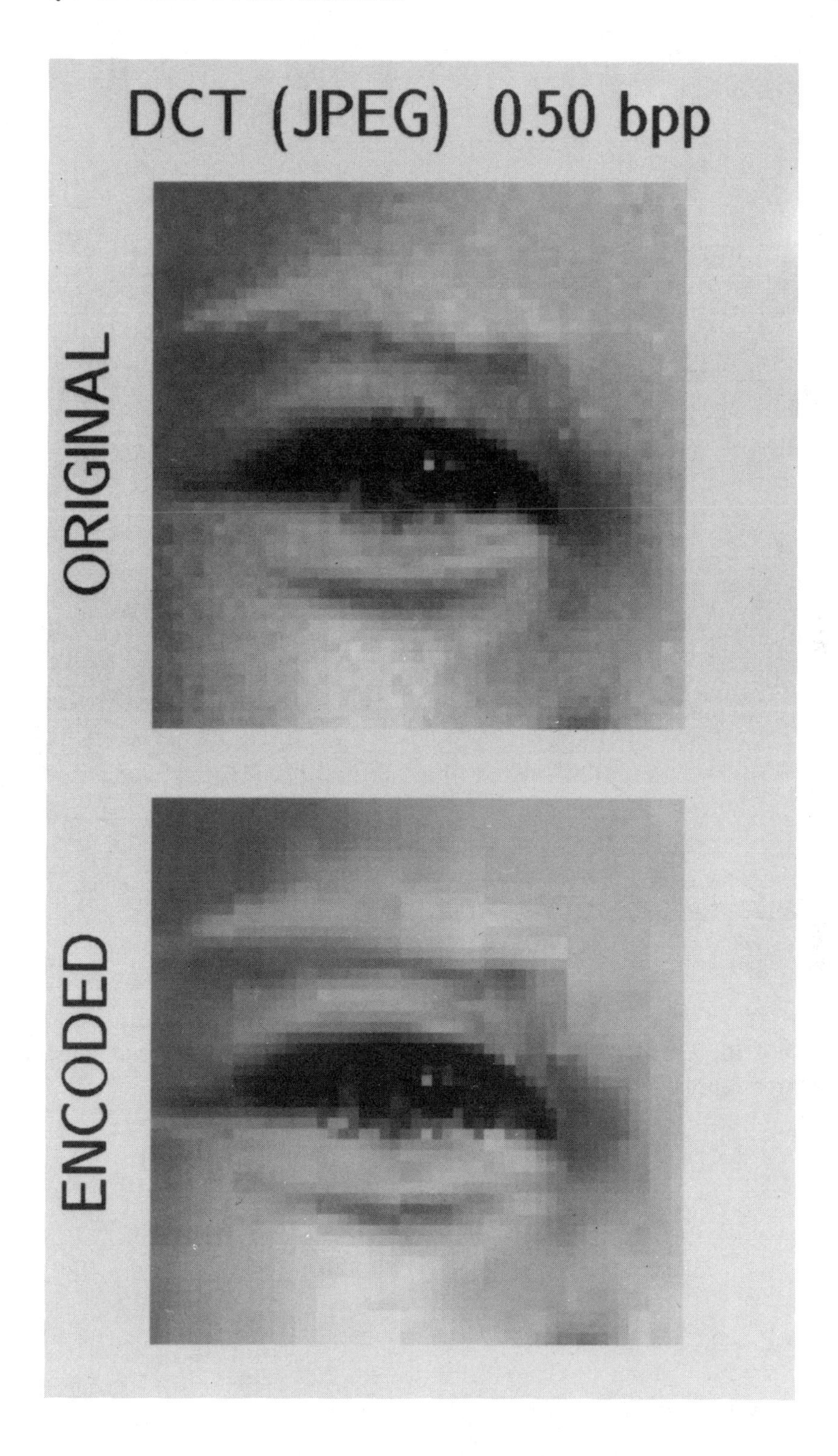

Figure 4.15: Enlarged portions of original and compressed images.

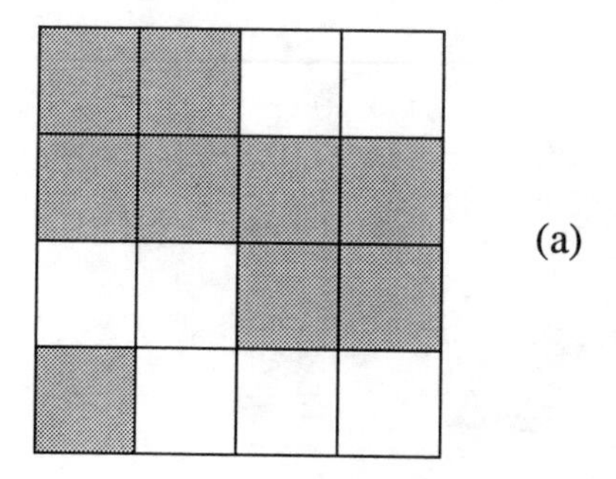

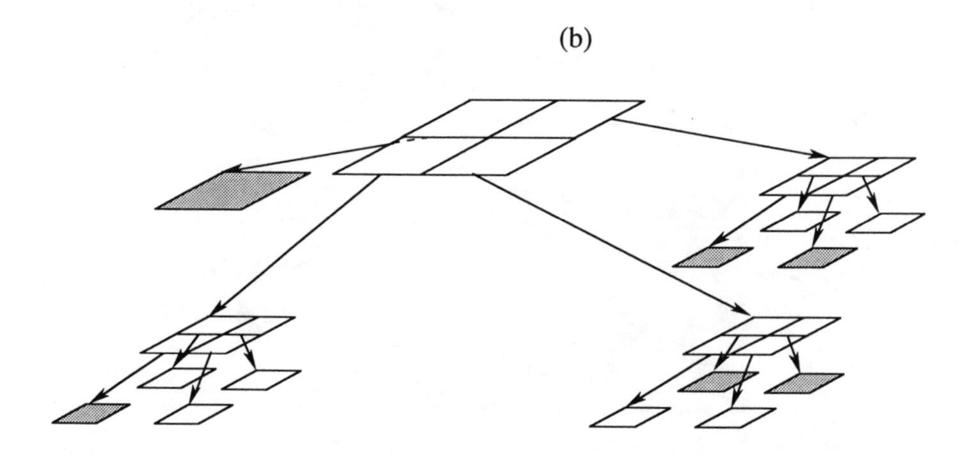

Figure 4.16: Quadtree decomposition of an image: (a) original image; (b) quadtree representation.

Chapter 5

Nonlinear Image Processing Algorithms

A key motivation for the study of nonlinear digital processing is the logical structure of the computer. All algorithms must be reduced to data and instruction flows through logic circuits and this requirement restricts the mathematical structures that are directly appropriate to digital processing. Linear filters involve addition and multiplication. Multiplications, when used in a linear convolution, also often involve real numbers (referred to as **floating point** numbers in computer architecture parlance), not just integers. Both addition and multiplication of floating point numbers take significantly longer than their integer counterparts since these operations must be implemented via a firmware subroutine involving multiple integer shifts, adds, and multiplies. Although some arithmetic logic units (ALUs) support floating point operations directly in microcode and coprocessors are sometimes added to facilitate floating point operations, these operations are still slower than integer operations.

As opposed to linear operations, nonlinear filters do not require multiplications. Although they may involve subtractions (additions), they only require integer operations. Perhaps the most salient point is that, for the most part, they involve maximum and minimum operations and their genesis lies in binary AND and OR operations, which are generally among the fastest macroinstructions provided by an ALU. We first study binary nonlinear filters defined via window logic.

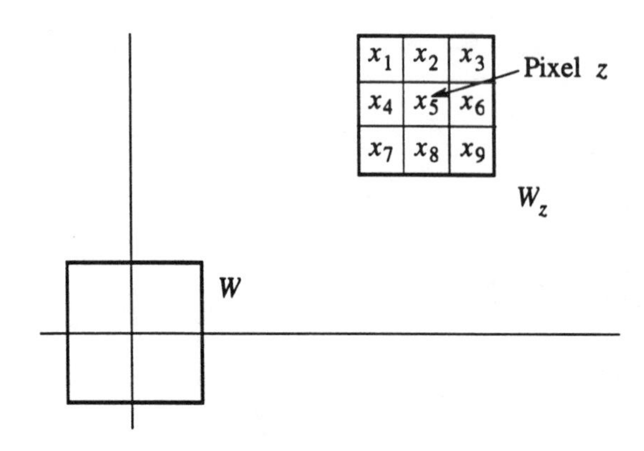

Figure 5.1: Translated logic mask.

5.1 Boolean Functions

A Boolean function is a binary function $h(x_1, x_2, \ldots, x_n)$ defined on n binary variables. Since each variable can take on two values, 0 or 1, there are 2^n possible arguments for h. In effect, h is defined by a truth table having 2^n rows: for each vector consisting of n 0s and 1s, there is associated a value of h, either 0 or 1. In conjunction with a window W, the Boolean function h defines a binary window operator Ψ on binary images via the one-to-one correspondence between the variables and pixels in the window. Ψ is defined at a pixel z by translating the window to z and applying the Boolean function h to the binary values in the translated window (see Fig. 5.1). The binary image is considered to be a set of pixels and the procedure is in accordance with the digitally stored image representation: 1 if a pixel is in the image and 0 if it is not. Geometrically, if A is the input image, then z is an element of A if and only if $A(z) = 1$, $A(z)$ denoting the value of A at z. Because the same Boolean function is applied for each translation of the window, the resulting window operator Ψ is translation (spatially) invariant.

From the standpoint of image processing, a binary window filter Ψ is defined by specifying the filter output for each possible configuration in the window. Thus, filter design involves synthesis of Boolean functions via truth tables. Given a truth table, construct an equivalent logical operator. Such a logical synthesis can always be accomplished directly from the truth values by defining the Boolean function in **disjunctive normal form**. Specifically, it can be represented by a logical sum of products (maximum of minima):

$$h(x_1, x_2, \ldots, x_n) = \sum_i x_1^{p(i,1)} x_2^{p(i,2)} \ldots x_n^{p(i,n)}, \tag{5.1}$$

where $p(i,j)$ is null or 'c,' indicating presence of the uncomplemented variable x_j or the complemented variable x_j^c in the ith product (**minterm**), respectively. There are up to 2^n minterms forming the maximum, one for each row of the corresponding truth table for which h has value 1. For instance, if there are three variables and the row 011 has value 1, then $x_1^c x_2 x_3$ is a minterm.

Reduction of the disjunctive normal form can be accomplished in accordance with the laws of logic to yield a reduced expansion

$$h(x_1, x_2, \ldots, x_n) = \sum_i x_{i,1}^{p(i,1)} x_{i,2}^{p(i,2)} \ldots x_{i,n(i)}^{p(i,n(i))}, \tag{5.2}$$

where $x_{i,1}$, $x_{i,2}, \ldots,$ $x_{i,n(i)}$ denote the $n(i)$ distinct variables in the ith product of the expansion and where there are at most 2^n distinct products in the expansion. Reduction is important because, if a binary filter is synthesized in its disjunctive normal form, that expression at once implies an implementation in combinational logic; however, reduction can provide a less costly implementation. Reduction is similarly beneficial for efficient software implementation. There are numerous reduction algorithms such as Karnaugh-map reduction, the Quine-McCluskey method, and the McCalla minterm-ring algorithm.

5.2 Increasing Boolean Functions

A Boolean function h is **increasing** if $(x_1, x_2, \ldots, x_n) \leq (y_1, y_2, \ldots, y_n)$ implies $h(x_1, x_2, \ldots, x_n) \leq h(y_1, y_2, \ldots, y_n)$, where $(x_1, x_2, \ldots, x_n) \leq (y_1, y_2, \ldots, y_n)$ if and only if $x_j \leq y_j$ for $j = 1, 2, \ldots, n$. Increasing Boolean functions are often called **positive** Boolean functions. A Boolean function h is positive if and only if it can be represented as a logical sum of products in which no variables are complemented:

$$h(x_1, x_2, \ldots, x_n) = \sum_i x_{i,1} x_{i,2} \cdots x_{i,n(i)}. \tag{5.3}$$

If the set of variables in any product of the expansion contains as a subset the set of variables in a distinct product, then, whenever the former product has value 1, so too does the latter. Thus, inclusion of the former product in the maximum expansion is redundant and can be deleted from the expansion without changing the function defined by h. No product whose variable set does not contain the variable set of a distinct product can be deleted without changing the function h. Performing the permitted deletions produces a **minimal representation** of the positive Boolean function. The majority of commonly employed nonlinear filters involve positive Boolean functions. We shall now introduce some key filters defined via Boolean functions.

Consider a single-product positive Boolean function

$$h_i(x_1, x_2, \ldots, x_n) = x_{i,1}x_{i,2}\ldots x_{i,n(i)}. \tag{5.4}$$

If the window operator Ψ is defined via h_i and A is the input image, then $z \in \Psi(A)$ if and only if $\Psi(A)(z) = 1$. In terms of window logic, this means that when the window W is translated to z, the pixels in the translated window corresponding to the pixels $w_{i,1}, w_{i,2}, \ldots, w_{i,n(i)}$ in W must all have value 1 so that the product of Eq. 5.4 is 1; that is, so $x_{i,1} = x_{i,2} = \ldots = x_{i,n(i)} = 1$. Let B denote the set of pixels $w_{i,1}, w_{i,2}, \ldots, w_{i,n(i)}$ in W. The pixels in the translated window corresponding to $w_{i,1}, w_{i,2}, \ldots, w_{i,n(i)}$ are $w_{i,1} + z, w_{i,2} + z, \ldots, w_{i,n(i)} + z$, and the product of Eq. 5.4 is 1 if and only if all of these pixel translates lie in the set A. The set of translates by z of pixels in B is denoted B_z. Thus, from a set perspective $z \in \Psi(A)$ if and only if B_z is a subset of A.

In image processing, the operation defined by the single logical product of Eq. 5.4 is called **erosion**. Since the logical product defined by Eq. 5.4 is fully defined by the pixel set B, the erosion Ψ is defined once B is specified. Erosion of binary image A by B is denoted by $A \ominus B$. In this context, B is called a **structuring element**.

Figure 5.2 shows a set A and a structuring element B (with the origin marked). Part (a) shows a translate of B to a pixel z for which the translate B_z is a subset of A, so that z lies in the filtered image $A \ominus B$. Part (b) shows a translate of B to a pixel w for which the translate B_w is not a subset of A, so that w does not lie in the filtered image. Part (c) shows $A \ominus B$.

Because positive Boolean functions possess noncomplemented logical representations of the kind given in Eq. 5.3, erosions form the building blocks of increasing binary nonlinear filters. Regarding the representation of Eq. 5.3, we can take a number of viewpoints. Relative to logical variables, it is a

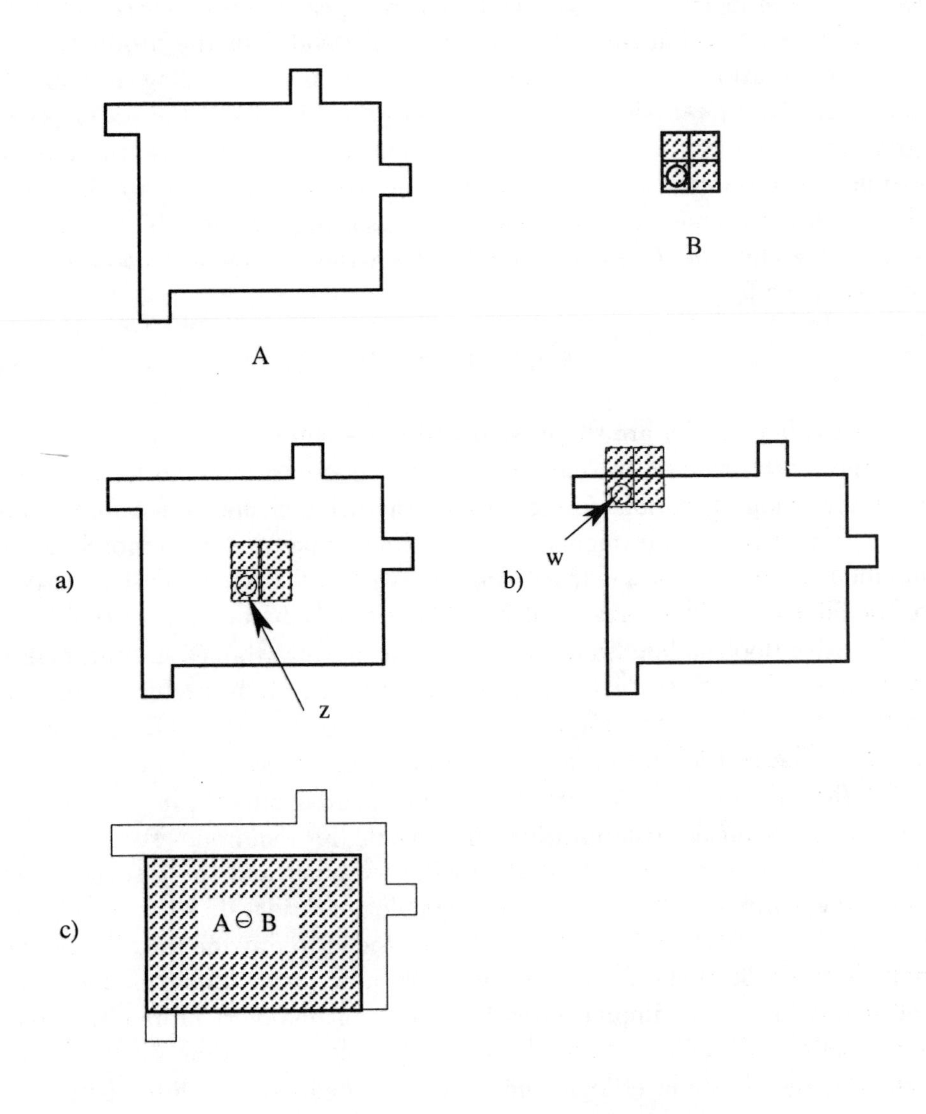

Figure 5.2: Erosion: (a) pixel in eroded set; (b) pixel not in eroded set; (c) eroded set.

logical sum of logical products. Architecturally, each erosion corresponds to an AND gate with an input for each logical variable in the product. The full representation is given by the various AND outputs feeding into an OR gate. Finally, we can view each erosion as providing a set, the set of pixels to which its structuring element can be translated and have the translate be a subset of the input image. The full filter, being defined by an OR function, then corresponds to the union of the resulting erosions. If there are m structuring elements (logical products), then the filter Ψ can be expressed by Eq. 5.3 or by

$$\Psi(A) = \bigcup_{i=1}^{m} A \ominus B_i, \tag{5.5}$$

where $B_1, B_2, \ldots, B_m$ are the m structuring elements.

Since each positive Boolean expression possesses a minimal form, so too does any union of erosions formed from structuring elements in a finite window. The structuring elements in the minimal expansion correspond to the minimal products. These structuring elements compose the basis, Bas $[\Psi]$, of the filter and the expansion of Eq. 5.5 need only be taken over the basis.

Positive Boolean functions are defined via logical variables and this makes them ideal for architectural analysis; nevertheless, it is useful to look at them in terms of fitting translates of structuring elements. This geometric approach lies at the root of morphological image processing.

Notice that if Eq. 5.5 were to be implemented directly in a high-level language, significant computation time would be required. For example, most programming languages that provide set types and the associated set operations of union, intersection, and complementation, do so at terrific execution time penalty – the data structures needed to provide the abstraction represent the bottleneck. Shortly we show how, using Boolean functions, set operations can be implemented in an efficient manner. Moreover, direct specification of nonlinear operations using Boolean notation allows for the construction of gate-level logic and hence fast, dedicated architectures.

There exists a morphological operation that is dual to erosion: **dilation** of set A by structuring element B is defined by

$$A \oplus B = (A^c \ominus (-B))^c, \tag{5.6}$$

where $-B = \{-b : b \in B\}$ is the reflection of B through the origin. Dilation also possesses an implementation without reference to complementation; however, we shall not pursue the matter here.

5.3 Nonlinear Noise Suppression Using Increasing Binary Filters

A key morphological filter is **opening**, which for a set A and structuring element B is defined by

$$A \circ B = \bigcup \{B_z : B_z \subset A\}. \tag{5.7}$$

$A \circ B$ is the union of all translates of B that are subsets of A: slide B around inside A and take as the filter output all pixels covered by the sliding structuring element. Whereas the position of the origin in the structuring element is crucial for erosion, it is irrelevant for opening. Opening can also be computed via erosion followed by dilation:

$$A \circ B = (A \ominus B) \oplus B. \tag{5.8}$$

If the observed image is an ideal image unioned with noise, then one way to employ opening is to choose a structuring element that fits inside the ideal image but not inside the noise. Opening will then remove the noise while keeping most of the signal. The method is depicted in Fig. 5.3 for a square degraded by background clutter. Opening restoration of a text image degraded by pepper noise is demonstrated in Fig. 5.4, which shows a text image, the text degraded by pepper, and the noisy image opened by a 3×3 square structuring element. It can be shown that opening is independent of structuring element location; thus the position of the origin need not be specified.

The positive Boolean function corresponding to opening by a 2×2 square structuring element is

$$h(x_1, x_2, \ldots, x_9) = x_1x_2x_4x_5 + x_2x_3x_5x_6 + x_4x_5x_7x_8 + x_5x_6x_8x_9, \tag{5.9}$$

where the 9 variables correspond to the pixels in the 3×3 square window centered at the origin (Fig. 5.1). The 2×2 opening can be found by moving the window and evaluating h at each location. The Boolean function can be implemented by the combinational logic circuit of Fig. 5.5 in which there are four AND gates and one OR gate, each having four inputs. An assembly program is implicit in Fig. 5.5. Morphologically, the filter is defined by Eq. 5.5 with $m = 4$ and the basis is composed of the structuring elements

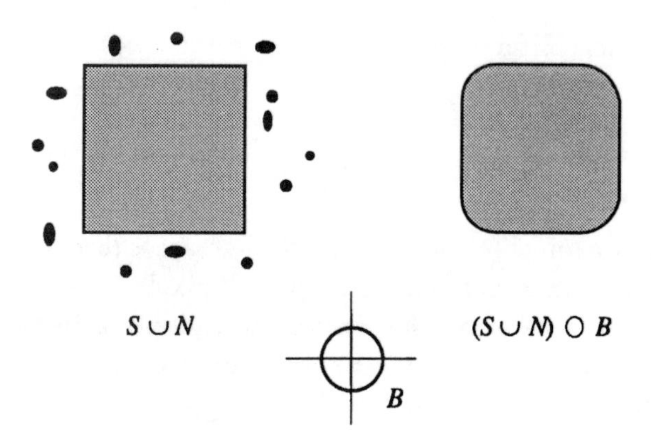

Figure 5.3: Suppression of clutter (union noise) by opening.

B_1, B_2, B_3, and B_4 given by

$$B_1 = \begin{bmatrix} 0 & 0 & 0 \\ 0 & \mathbf{1} & 1 \\ 0 & 1 & 1 \end{bmatrix} \quad B_2 = \begin{bmatrix} 0 & 1 & 1 \\ 0 & \mathbf{1} & 1 \\ 0 & 0 & 0 \end{bmatrix}$$
$$B_3 = \begin{bmatrix} 0 & 0 & 0 \\ 1 & \mathbf{1} & 0 \\ 1 & 1 & 0 \end{bmatrix} \quad B_4 = \begin{bmatrix} 1 & 1 & 0 \\ 1 & \mathbf{1} & 0 \\ 0 & 0 & 0 \end{bmatrix}. \tag{5.10}$$

The union expansion is graphically illustrated in Fig. 5.6 for the set A of Fig. 5.2.

A commonly employed filter for restoring binary images is the median filter. Given a window W containing an odd number of pixels, say n, the binary **moving median** is defined in the following manner: for each pixel z, W is translated to z and the filter outputs 1 if more than $n/2$ pixels in W_z are 1-valued; otherwise, the filter outputs 0. Medians are used to suppress salt-and-pepper noise and exhibit good edge preservation if the amount of noise is not excessive and the uncorrupted image does not possess much fine detail. Figures 5.7 through 5.10 show salt-and-pepper-degraded and median-restored text images. Notice how the 5×5 median has yielded good

(a)

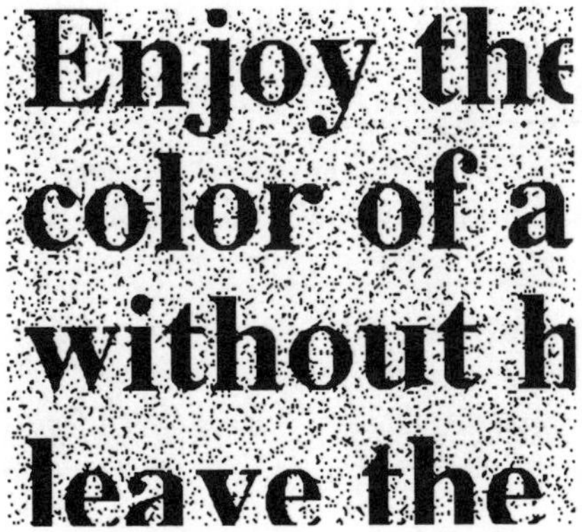

(b)

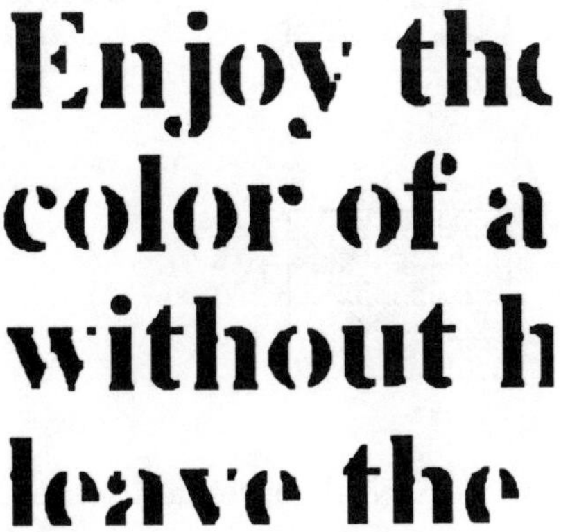

(c)

Figure 5.4: Restoration of text by opening: (a) original image, (b) image degraded by pepper noise, (c) filtered image.

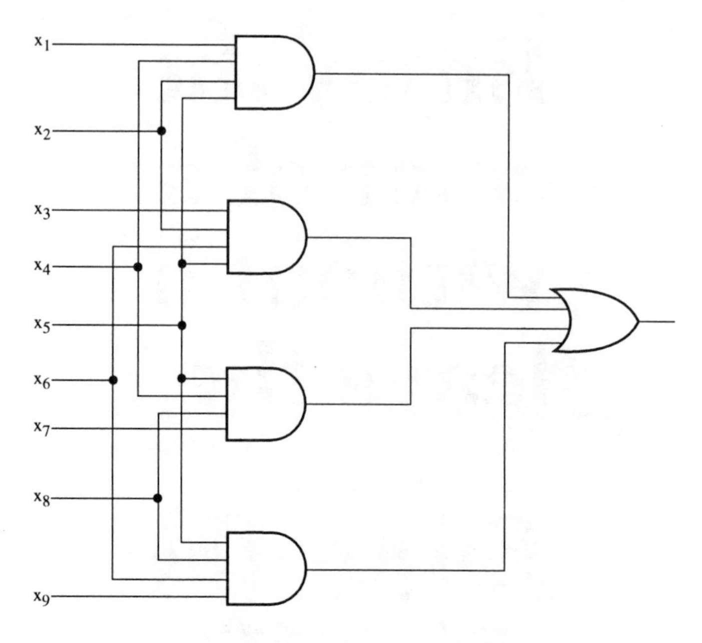

Figure 5.5: Combinational logic for 2 × 2 opening.

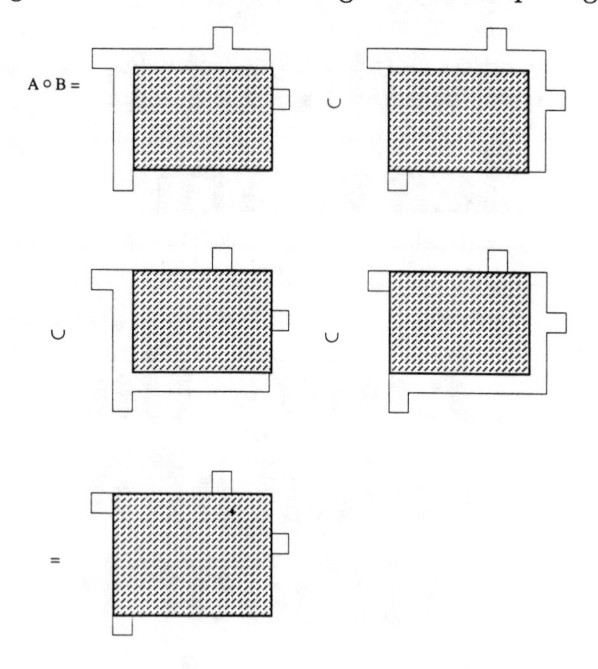

Figure 5.6: Morphological representation of 2 × 2 opening.

restoration for the simple Helvetica font with only 5% noise in Fig. 5.7, but poorer quality restoration for 20% noise and the finely detailed Monotype Corsiva font.

Because the median is based on pixel counts in windows, it is an increasing operator and must possess a basis expansion according to Eq. 5.5. Consider the median over the **strong-neighbor mask** (the origin and the pixels immediately above, below, left, and right of it). A pixel lies in the filtered image if and only if, among itself and its four strong neighbors, at least three pixels lie in the input image. The following ten structuring elements compose a basis:

$$\begin{bmatrix} 0 & 1 & 0 \\ 1 & \mathbf{1} & 0 \\ 0 & 0 & 0 \end{bmatrix} \begin{bmatrix} 0 & 1 & 0 \\ 1 & \mathbf{0} & 1 \\ 0 & 0 & 0 \end{bmatrix} \begin{bmatrix} 0 & 1 & 0 \\ 0 & \mathbf{1} & 1 \\ 0 & 0 & 0 \end{bmatrix} \begin{bmatrix} 0 & 1 & 0 \\ 1 & \mathbf{0} & 0 \\ 0 & 1 & 0 \end{bmatrix}$$

$$\begin{bmatrix} 0 & 1 & 0 \\ 0 & \mathbf{1} & 0 \\ 0 & 1 & 0 \end{bmatrix} \begin{bmatrix} 0 & 1 & 0 \\ 0 & \mathbf{0} & 1 \\ 0 & 1 & 0 \end{bmatrix} \begin{bmatrix} 0 & 0 & 0 \\ 1 & \mathbf{1} & 1 \\ 0 & 0 & 0 \end{bmatrix} \begin{bmatrix} 0 & 0 & 0 \\ 1 & \mathbf{1} & 0 \\ 0 & 1 & 0 \end{bmatrix} \tag{5.11}$$

$$\begin{bmatrix} 0 & 0 & 0 \\ 1 & \mathbf{0} & 1 \\ 0 & 1 & 0 \end{bmatrix} \begin{bmatrix} 0 & 0 & 0 \\ 0 & \mathbf{1} & 1 \\ 0 & 1 & 0 \end{bmatrix}.$$

The strong-neighbor median can be evaluated by eroding by each of the ten structuring elements and then forming the union of Eq. 5.5.

To obtain a logical expression of the strong-neighbor median according to Eq. 5.3, we assign five variables to the window pixels to obtain the logical mask shown in Fig. 5.11. The Boolean function corresponding to the union of erosions with the 10 structuring elements of Eq. 5.11 is

$$\begin{aligned} h &= x_1x_2x_3 + x_1x_2x_4 + x_1x_3x_4 + x_1x_2x_5 + x_1x_3x_5 \\ &\quad + x_1x_4x_5 + x_2x_3x_4 + x_2x_3x_5 + x_2x_4x_5 + x_3x_4x_5. \end{aligned} \tag{5.12}$$

Both a hardware implementation and an assembly program are implicit in Eq. 5.13.

We next illustrate how nonincreasing binary filters can be used for restoration. First we give the morphological interpretation of the general representations of Eqs. 5.1 and 5.2.

Consider a window W and two disjoint subsets E and F of W. A filter Ψ is defined in the following manner: if A is an input image and z is any

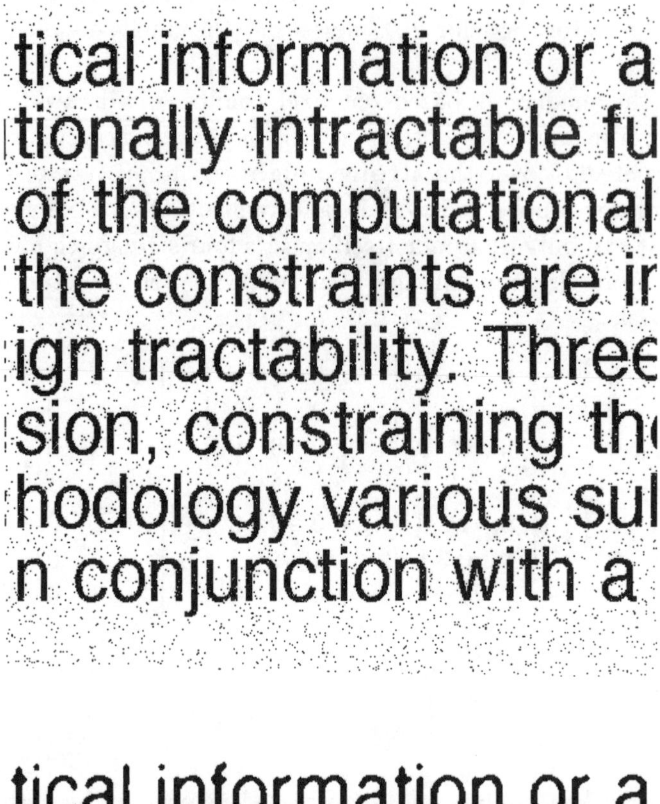

(a)

(b)

Figure 5.7: Median restoration of Helvetica text degraded by 5% salt-and-pepper noise: (a) degraded image; (b) restored image.

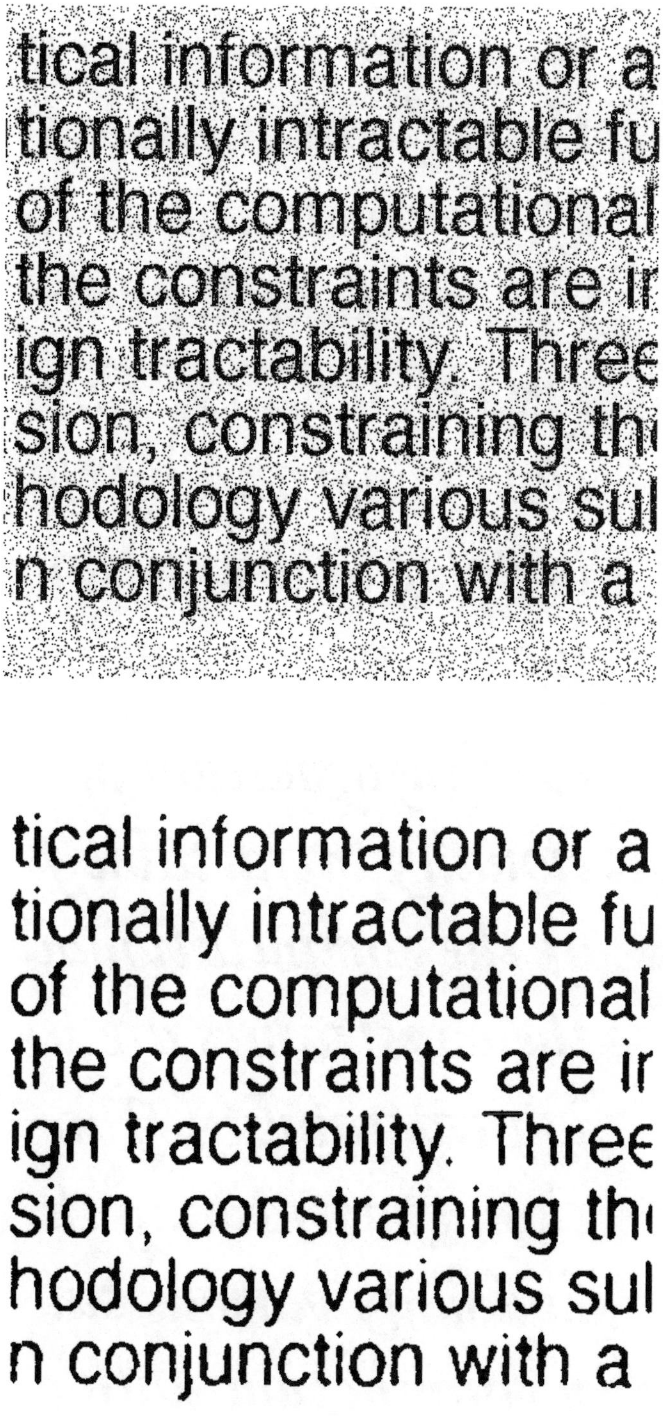

(a)

(b)

Figure 5.8: Median restoration of Helvetica text degraded by 20% salt-and-pepper noise: (a) degraded image; (b) restored image.

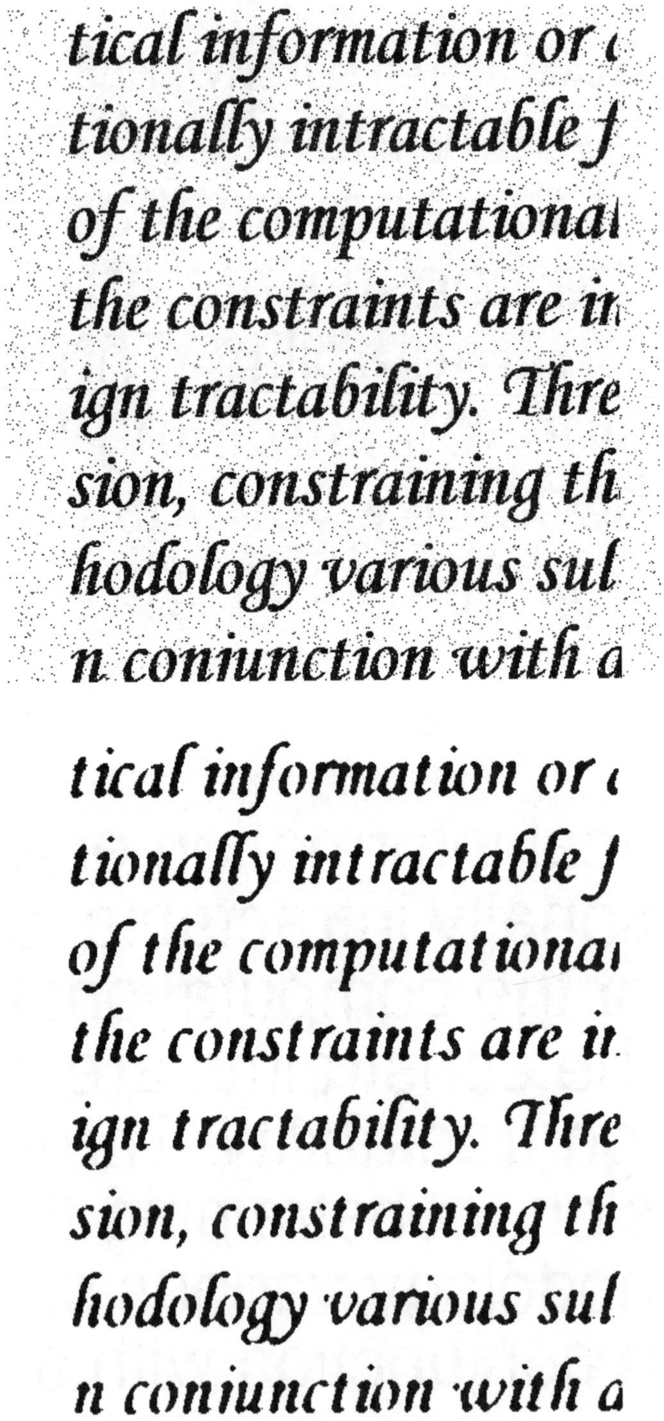

tical information or
tionally intractable
of the computational
the constraints are in
ign tractability. Thre
sion, constraining th
hodology various sul
n coniunction with a

(a)

tical information or
tionally intractable
of the computational
the constraints are in
ign tractability. Thre
sion, constraining th
hodology various sul
n coniunction with a

(b)

Figure 5.9: Median restoration of Monotype Corsiva text degraded by 5% salt-and-pepper noise: (a) degraded image; (b) restored image.

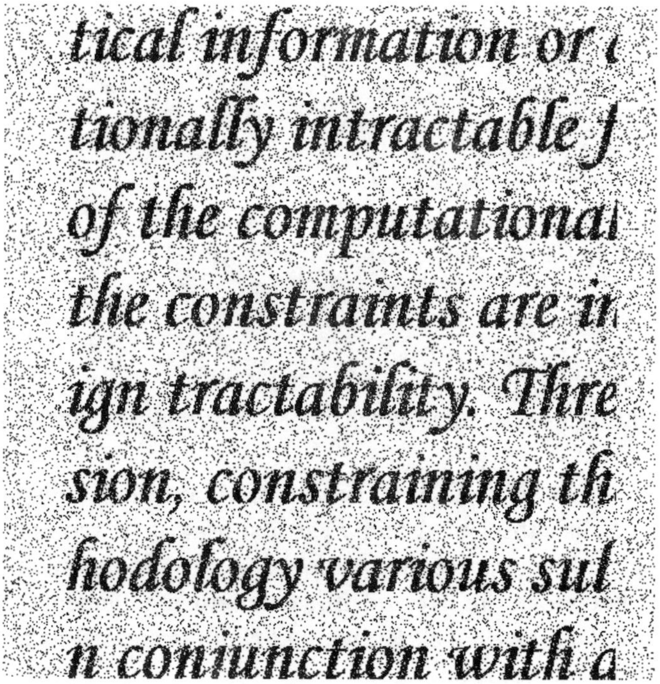

tical information or
tionally intractable
of the computational
the constraints are in
ign tractability. Thre
sion, constraining th
hodology various sul
n coniunction with a

(a)

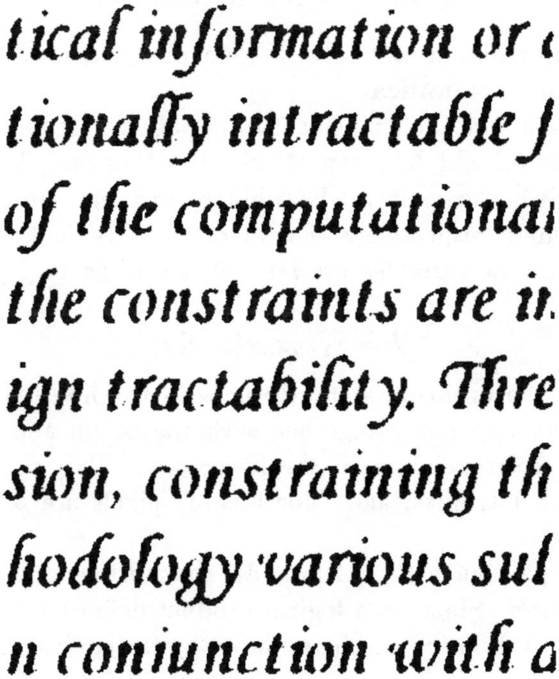

tical information or
tionally intractable
of the computational
the constraints are in
ign tractability. Thre
sion, constraining th
hodology various sul
n coniunction with a

(b)

Figure 5.10: Median restoration of Monotype Corsiva text degraded by 20% salt-and-pepper noise: (a) degraded image; (b) restored image.

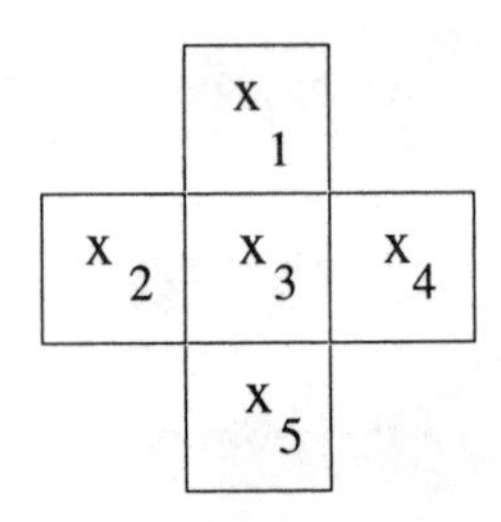

Figure 5.11: Logical mask for strong-neighbor median.

pixel, then $z \in \Psi(A)$ if E_z is a subset of A and F_z is disjoint from A. Equivalently, E_z is subset of A and F_z is a subset of A^c (complement of A). In morphological image processing, Ψ is called the **hit-or-miss transform,** E and F are known as the **hit** and **miss** structuring elements, respectively, and $\Psi(A)$ is written as $A \circledast (E, F)$, where (E, F) is treated as a structuring-element pair. In set notation,

$$A \circledast (E, F) = (A \ominus E) \cap (A^c \ominus F). \tag{5.13}$$

If E and F form a decomposition of W, meaning $E \cup F = W$, then the pair (E, F) is said to be **canonical**.

Canonical and noncanoncial structuring pairs correspond to single product terms in Eqs. 5.1 and 5.2, respectively. In either case, E and F represent the sets of pixels in the window whose binary variables are uncomplemented and complemented, respectively. For instance, if W is 3×3 and centered at the origin and its variables are labeled according to Fig. 5.1, then the product

$$h = x_1 x_2 x_3 x_5 x_7^c x_8^c x_9^c \tag{5.14}$$

corresponds to the hit-or-miss transform with the hit structuring element consisting of the upper row together with the origin and the miss structuring element corresponding to the lower row. This pair is depicted in Fig. 5.12, where hit, miss, and "don't care" pixels are black, white, and gray, respectively.

The logic expansion of Eq. 5.1 provides the most general form of a binary windowed operator. Since each logical product defines a hit-or-miss operator, the function expansion can be replaced by the equivalent morphological

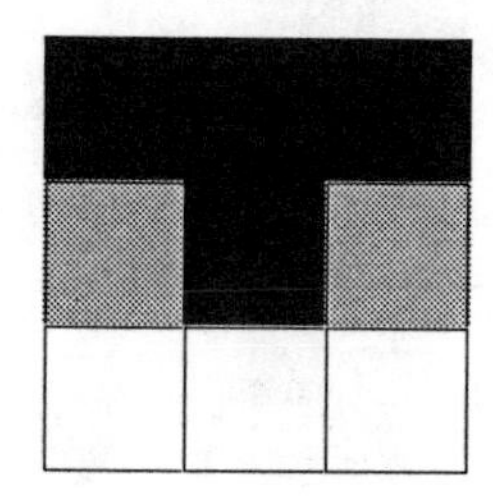

Figure 5.12: Hit-or-miss structuring pair.

expansion

$$\Psi(A) = \bigcup_i A \circledast (E_i, F_i), \tag{5.15}$$

where (E_i, F_i) is the structuring element pair corresponding to the ith logical product.

5.4 Noise Suppression Using Nonincreasing Binary Filters

The hit-or-miss transform is employed in various ways to restore noise-degraded images. We illustrate only one approach, **parallel thinning.** Here, the original image has been degraded by adjoining noise pixels. Hit-or-miss transforms are used to identify pixels to be removed and then these are subtracted from the image. The resulting thinning filter takes the form

$$\Psi(A) = A - \bigcup_i A \circledast (E_i, F_i), \tag{5.16}$$

where the pairs (E_i, F_i) are designed to locate noise pixels.

A standard approach to thinning is to employ the eight pruning structuring pairs shown in Fig. 5.13. Examination of the pruners reveals them to be endpoint finders for the eight principal directions in the square grid. The assumption in using them for thinning is that most endpoints in an image degraded by spurious adjoined edge pixels are noise pixels. The pruners find endpoints and then, according to Eq. 5.16, the endpoints are removed

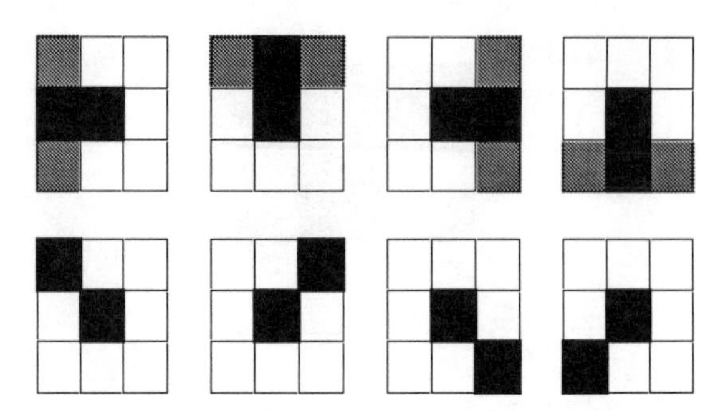

Figure 5.13: Classical pruners.

to yield the filtered image. Figure 5.14 shows an original text image, an edge-degraded version of the text image, and the result of applying parallel thinning to the degraded image using the eight pruners.

The classical pruners work well for noncomplex noise; however, for realistic noise models it is better to design the structuring pairs in a statistically optimal manner. A similar comment applies to application of increasing binary filters.

Implementation of Eq. 5.16 directly in a high-level language will usually result in very inefficient code, especially for large images.

5.5 Matched Filtering Using Nonincreasing Binary Filters

From its definition, the hit-or-miss transform is a matched filter, the hit structuring element fitting inside and the miss element fitting outside. The difficulty is that, for matched filtering, the window must be large: it must be sufficiently large to encompass whatever shapes are to be identified. Moreover, each element might contain a large number of pixels and therefore require a large amount of processing.

To illustrate the method, consider the letter "A" shown at the extreme left of the upper row in Fig. 5.15 and its complement image shown at the extreme left of the second row. The second image in the top row is an inner boundary for "A" and succeeding images to the right show increasingly collapsed versions. The second image in the second row is an outer boundary

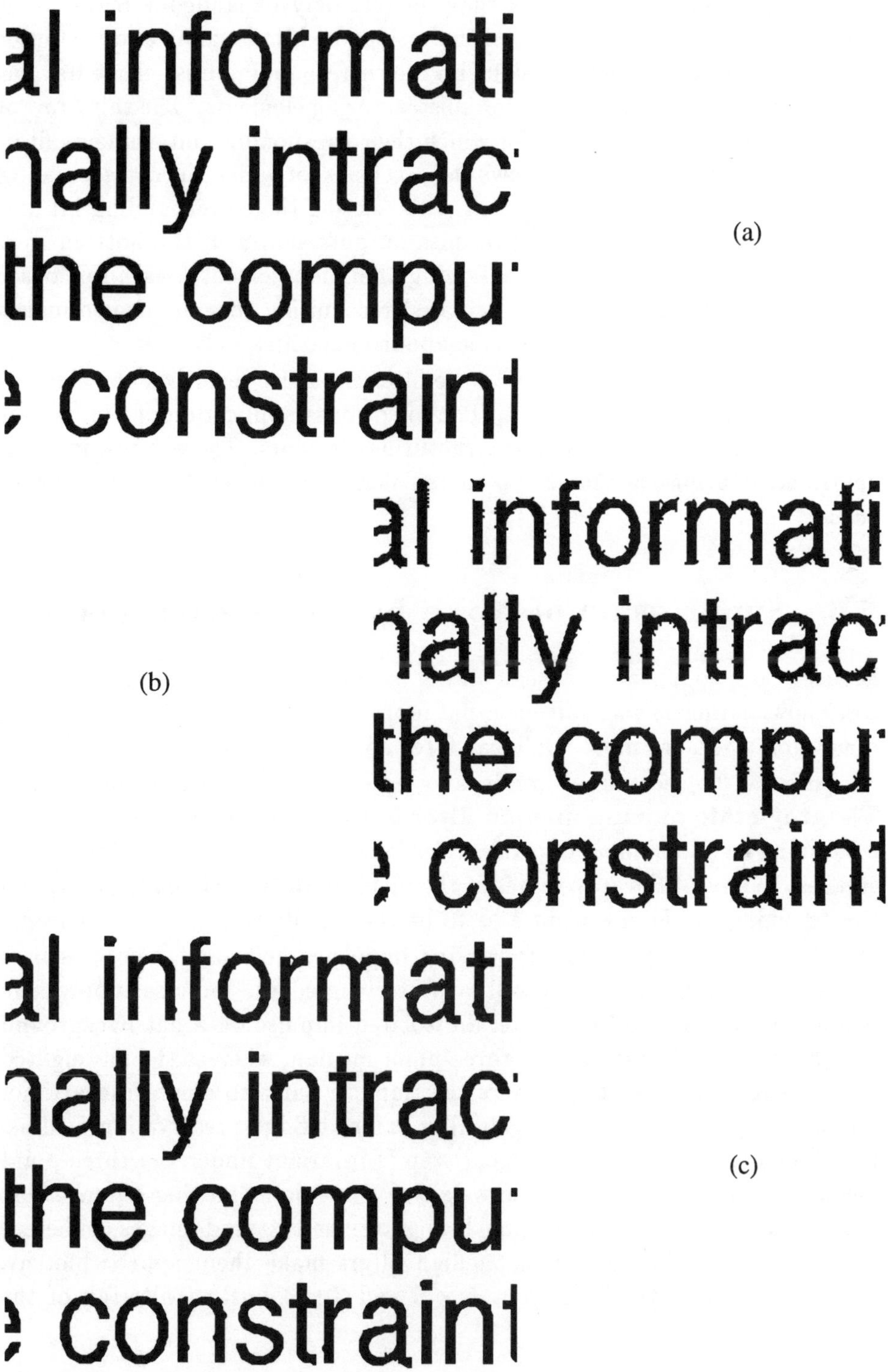

(a)

(b)

(c)

Figure 5.14: Text restoration using parallel thinning: (a) original text image; (b) edge-degraded text image; (c) restored text.

for "A" and succeeding images show increasingly expanded versions. Let the inner boundary and its shrunken versions be hit elements and the outer boundary and its enlarged versions be the corresponding miss elements. Let the window center be the origin for all structuring elements. The third row of the figure shows the pixel sets for which the corresponding hit elements fit in the letter and the fourth row shows the pixel sets for which the corresponding miss elements fit in the complement. Intersecting the corresponding hit and miss pixel sets provides the hit-or-miss outputs shown in the bottom row. Shrinking the hit element and enlarging the miss element makes for looser fitting. Loose fitting results in increased recognition regions; it also makes the matched filter less sensitive to shape irregularities.

To reduce processing and make the filter less noise sensitive, one can employ sparse structuring elements. The hit-or-miss application of Fig. 5.15 is repeated in Fig. 5.16 with sparse structuring elements. There is risk in using sparse structuring elements: too sparse elements can result in misrecognitions.

5.6 Suppression of Noise by Gray-Scale Median

In discussing nonlinear gray-scale filters we focus on a class of increasing filters known either as stack filters or flat morphological filters. This restriction is not great because most nonlinear-filter applications involve this class.

Perhaps the most used nonlinear gray-scale filter is the median filter. The **gray-scale moving median filter** is defined by choosing a window W containing an odd number of pixels, translating W to a pixel z, ordering from smallest to largest the gray values at the pixels in the translated window, and then defining the filter output at z to be the middle value in the ordering.

The median is good for suppressing impulsive noise and, as opposed to linear smoothing filters, does well at preserving edges. Impulse suppression is illustrated in Fig. 5.17. There, an isolated impulse on a flat background signal is fully suppressed by the three-point median, whereas the unweighted three-point moving average lowers the impulse but also creates an artifact around it by raising some neighboring values. Edge preservation is illustrated in Fig. 5.18. There, a signal step is invariant under the three-point median but is ramped by the unweighted three-point moving average. As with linear smoothing, the median does poorly on textured images. Whereas linear smoothers blur textures, median filters make them appear blotchy. Figures 5.19 through 5.22 show 5×5 and 3×3 median filtering of the

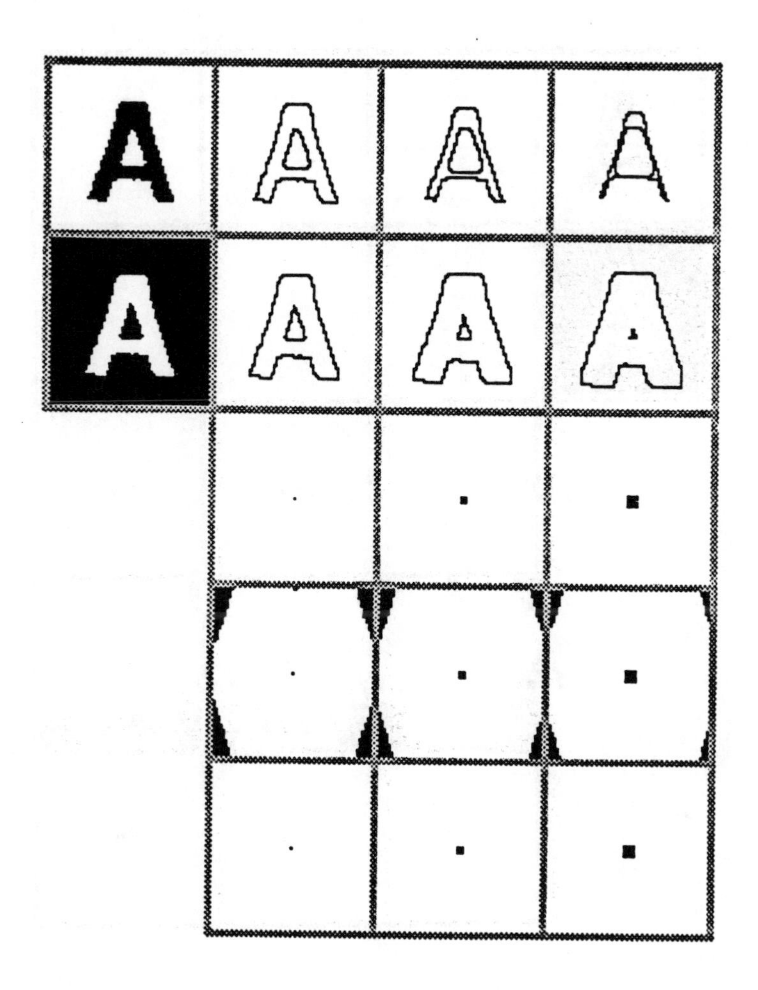

Figure 5.15: Hit-or-miss recognition.

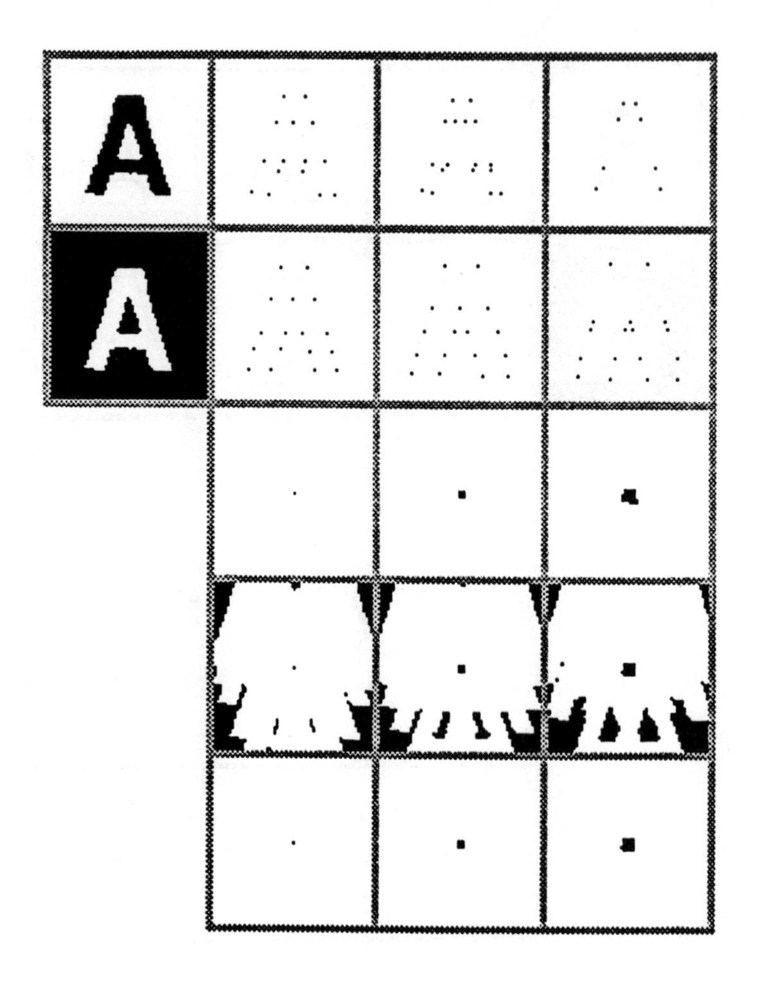

Figure 5.16: Hit-or-miss recognition with sparse structuring elements.

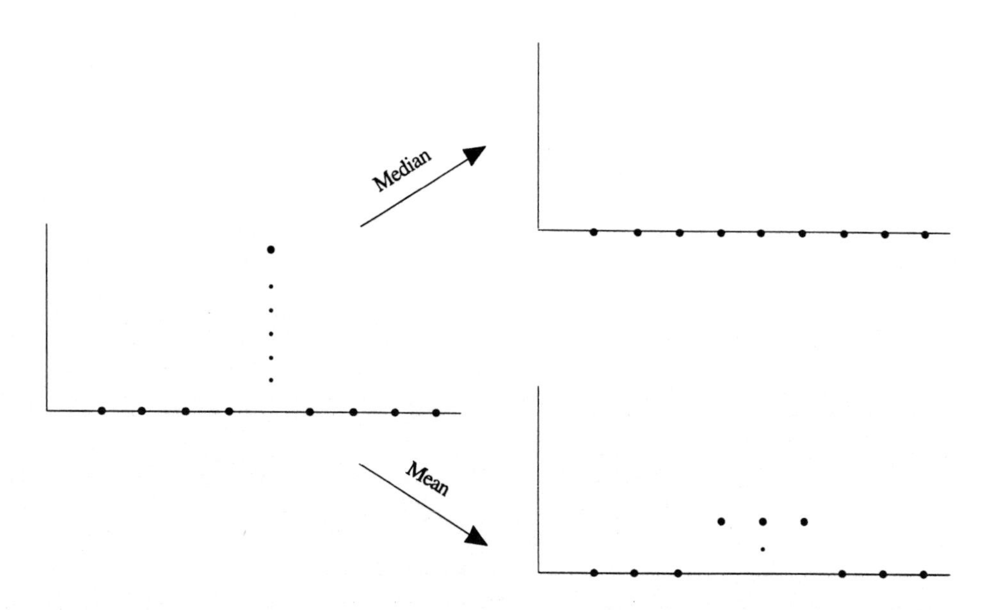

Figure 5.17: Impulse responses for three-point median and moving mean.

impulse-degraded images of Figs. 3.2 through 3.5, respectively. Notice the more pronounced edges produced by the median filter than the moving mean and the blotchy effect of the median on the textured portions of the mandrill.

For linear filtering, if one wishes to give more weight to the center pixel or to pixels near the center, the convolution can be so weighted; a similar concept applies to medians. Suppose $x_1, x_2, \ldots, x_m$ are the observed values in the structuring set. The **weighted median** with integer weights $a_1, a_2, \ldots, a_m$ is found by repeating a_i times the observation x_i, ordering the new set of $a_1 + a_2 + \ldots + a_m$ values, and then choosing the middle value as the output. In particular, the **center weighted median** is found by repeating only the value at the window center.

5.7 Threshold Decomposition

The median is a special case of a more general kind of increasing nonlinear gray-scale filter, these being defined via the method of threshold decomposition. Suppose the gray range has $M + 1$ values, $0, 1, \ldots, M - 1, M$. For

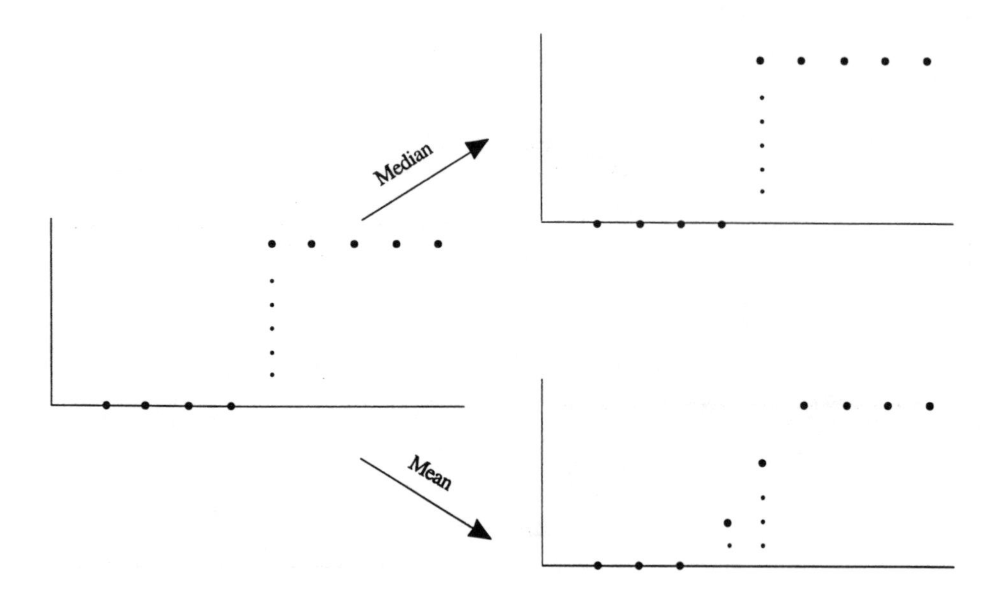

Figure 5.18: Step responses for three-point median and mean.

each gray value k, define the **threshold set** of the image f by

$$A[k] = \{z : f(z) \geq k\} \tag{5.17}$$

(Fig. 5.23). If $k \leq j$, then $A[j]$ is a subset of $A[k]$. Thus, if an increasing binary filter Ψ_0 is applied to both $A[k]$ and $A[j]$, then $\Psi_0(A[j])$ is a subset of $\Psi_0(A[k])$. One can imagine the threshold sets of f as stacking on top of each other, with each being a subset of the one below it.

If an increasing binary filter Ψ_0 is applied to each set in the stack, then the stack sets are changed, but the order relations are preserved. A gray-scale filter Ψ is defined by taking $\Psi(f)(z)$, the value of the filtered signal at pixel z, to be the maximum gray value k for which z is an element of $\Psi_0(A[k])$. Equivalently,

$$\Psi(f)(z) = \sum_{k=1}^{M} \Psi_0(A[k])(z). \tag{5.18}$$

The resulting filter Ψ is called a **stack filter**. Since Ψ_0 is an increasing binary filter, it is generated by a positive Boolean function, so that, in effect, at each level of the stack the same logical operator (of the form given in Eq. 5.3) is being applied to the variables in the defining window about each pixel. The

(a)

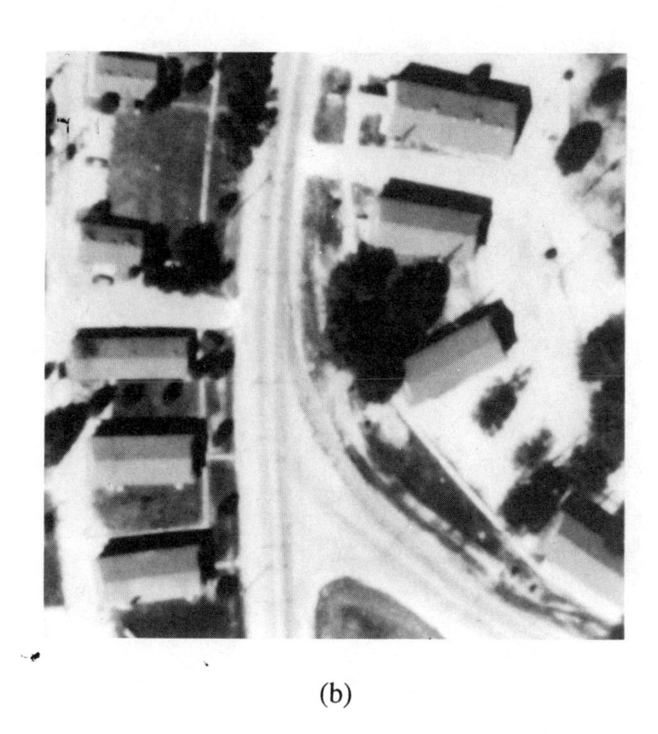

(b)

Figure 5.19: Median filtering of aerial scene: (a) 5×5; (b) 3×3.

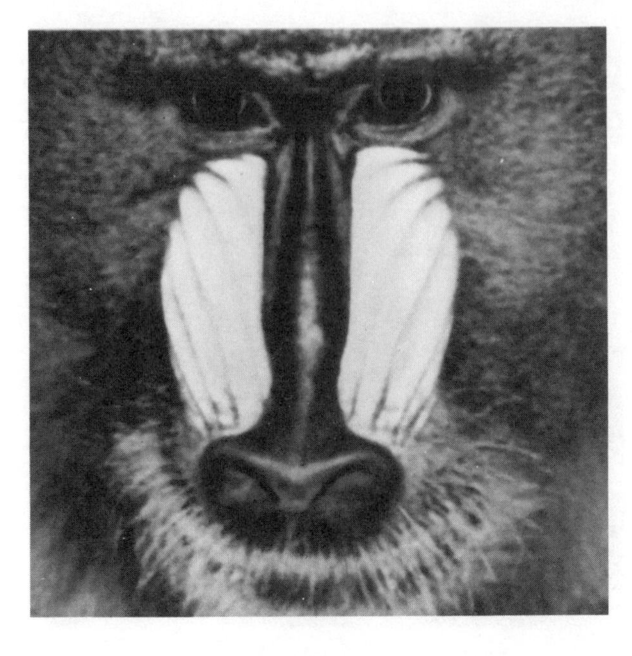

(a)

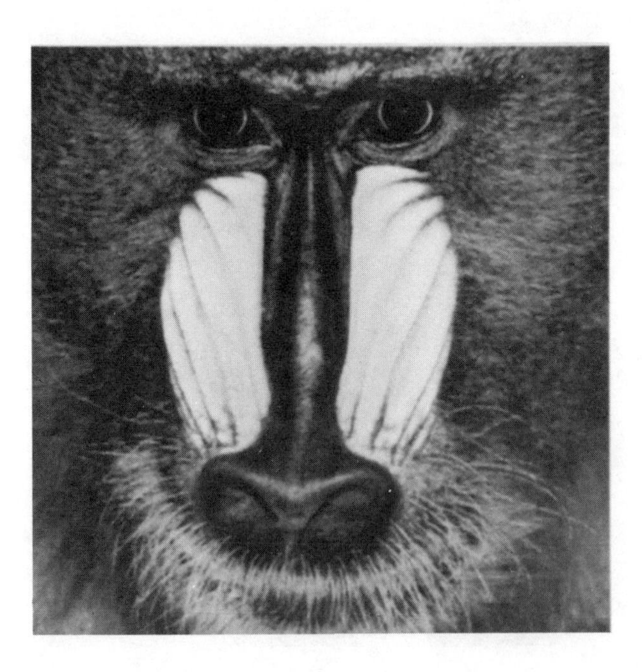

(b)

Figure 5.20: Median filtering of mandrill: (a) 5×5; (b) 3×3.

(a)

(b)

Figure 5.21: Median filtering of bridge scene: (a) 5×5; (b) 3×3.

(a)

(b)

Figure 5.22: Median filtering of house scene: (a) 5×5; (b) 3×3.

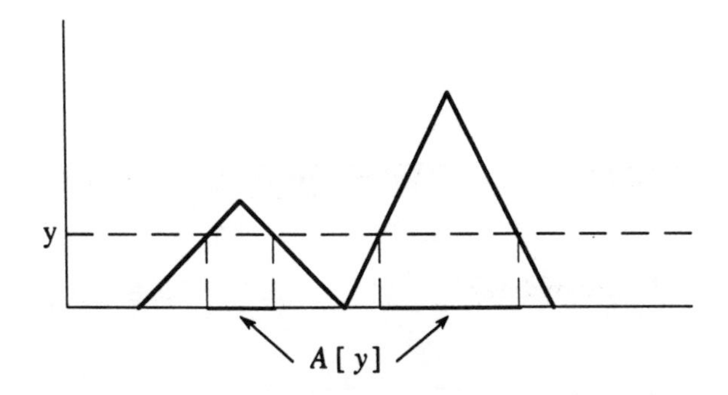

Figure 5.23: Threshold set.

method is illustrated for the three-point centered median in Fig. 5.24. On the left are the threshold sets of the input signal and on the right are the threshold sets of the output signal, each output threshold set having been obtained by applying the binary median (the positive Boolean function for the binary median) to the corresponding input threshold set. Architecturally and programatically, a gray-scale stack filter is built using only a single binary logical function.

Stack filters can be viewed morphologically. If f is an image and E is a set, then the **moving minimum** over E is found by translating E to a pixel z and then taking the minimum of all values in the translate E_z. The **moving maximum** is defined analogously. In the context of morphological image processing, the moving minimum is called a **flat erosion**, the moving maximum is called a **flat dilation**, and E is the structuring element. Flat erosion and flat dilation are denoted by $f \ominus E$ and $f \oplus E$, respectively. Both flat erosion and dilation are notoriously slow running if coded directly in a high-level language for a SISD processor. We recommend always coding these operations using their Boolean analogs in both high-level languages and in assembly language.

Suppose h is a positive Boolean function defining stack filter Ψ, h has the logical sum-of-products representation given in Eq. 5.3, and h is applied over the variables in the window W. Each product in Eq. 5.3 corresponds to a subset E_i of W and Ψ has the maximum representation

$$\Psi(f) = \bigvee_i f \ominus E_i. \tag{5.19}$$

In short, a stack filter is a maximum of flat erosions.

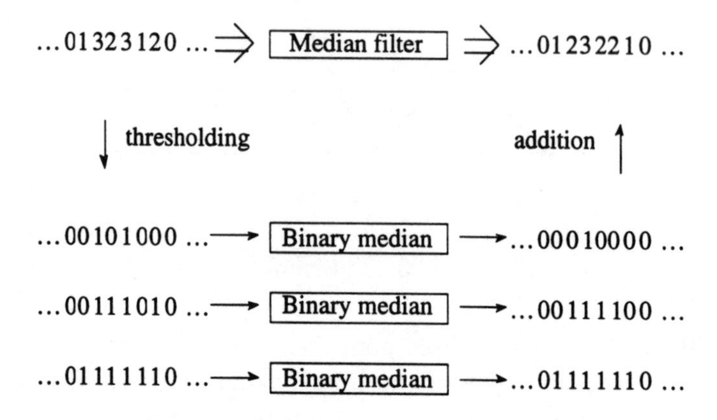

Figure 5.24: Median as a stack filter.

Regarding the median, the structuring elements of Eq. 5.11, which served as a basis for the binary moving strong-neighbor median, also serve as the structuring elements for the gray-scale moving strong-neighbor median; that is, if 10 erosions are employed in Eq. 5.19 using the structuring elements of Eq. 5.11, then the resulting filter Ψ is the strong-neighbor median.

5.8 Nonlinear Edge Detection Via the Morphological Gradient

The nonlinear analog to linear gradient edge detection is edge detection via the morphological gradient. Given an image f and a structuring element E, the **morphological gradient** is defined by

$$\mathrm{GRAD}[f] = [f \oplus E] - [f \ominus E], \tag{5.20}$$

where $-$ denotes the set difference. Because dilation and erosion by sets yield maximum and minimum filters, respectively, at each pixel the morphological gradient yields the difference between the maximum and minimum values over the set translated to the pixel. Its action is illustrated in Fig. 5.25. Like linear gradients, once the morphological gradient has been found, an

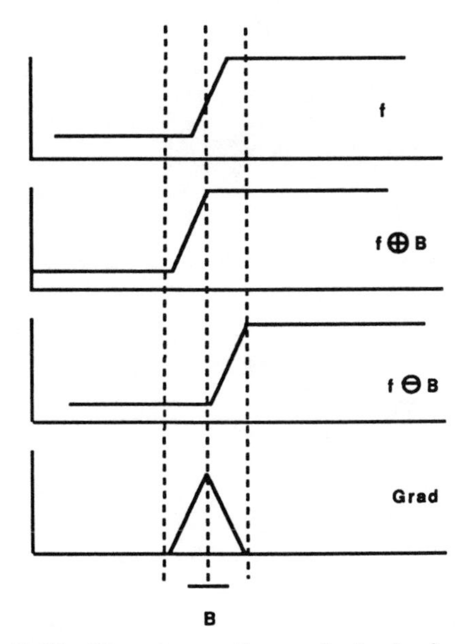

Figure 5.25: Structure of morphological gradient.

edge image can be created by thresholding. The morphological gradient followed by thresholding is illustrated in Figs. 5.26 and 5.27 for the aerial and house scenes of Figs. 3.2 and 3.5, respectively. Rather than use gradient edge detection as a final edge result for image segmentation, it is better to use it as part of a more accurate morphological method known as **watershed segmentation**.

5.9 Fast Granulometric Filters

Fast matrix transforms reduce computation time by providing equivalent implementations involving sparse matrices containing mostly zeros and ones. More generally, one often attempts to find an equivalent algorithmic implementation that saves processing time on the available hardware. In the present section we consider a nonlinear algorithm that lies at the base of a very powerful texture discrimination methodology and discuss a fast implementation.

The opening filter introduced in Section 5.3 has a number of key properties, including the fact that the opening $A \circ B$ is always a subset of A. Moreover, for certain sequences of structuring elements that are increasing

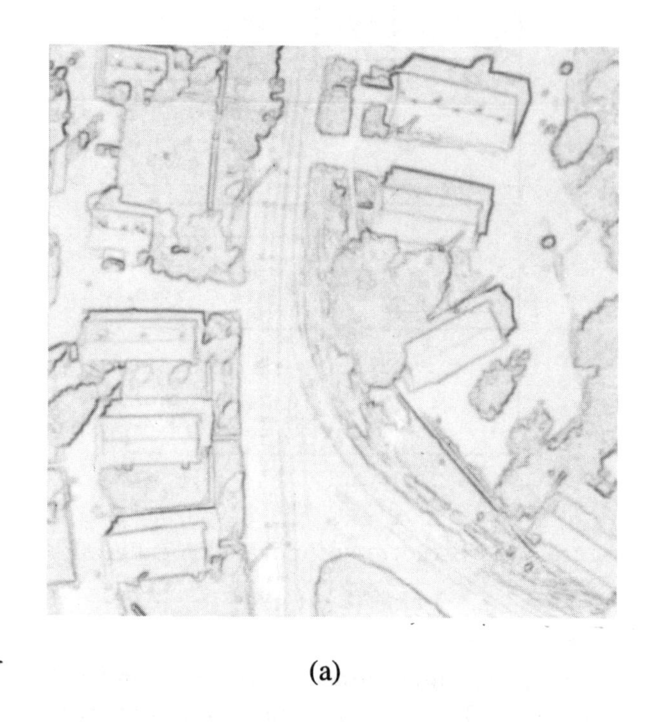

(a)

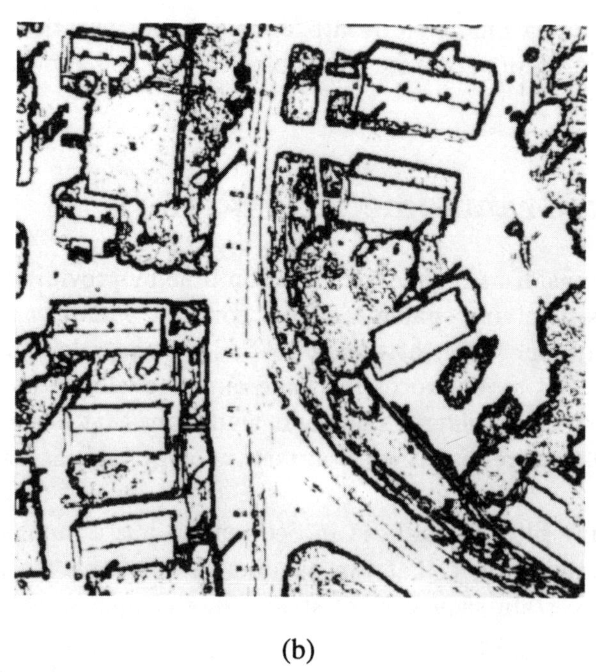

(b)

Figure 5.26: Morphological gradient applied to aerial scene: (a) gradient image; (b) edge image.

(b)

Figure 5.27: Morphological gradient applied to house scene: (a) gradient image; (b) edge image.

in size, say $B_0 \subset B_1 \subset B_2 \ldots$, the corresponding openings are decreasing in size, namely, $A \circ B_0 \supset A \circ B_1 \supset A \circ B_2 \supset \ldots$. The sequence of openings is called a **granulometry.** A **granulometric size distribution** $\Omega(k)$ corresponding to the binary image A is generated by counting the pixels in each succeeding filtered image $A \circ B_k$. $\Omega(k)$ is a decreasing function of k and, for sufficiently large k, $\Omega(k) = 0$. If we assume that B_0 consists of a single pixel, then $\Omega(0)$ gives the original pixel count in A and the normalization

$$\Phi(k) = 1 - \frac{\Omega(k)}{\Omega(0)} \tag{5.21}$$

is a digital function that increases from 0 to 1 over some finite k-interval. $\Phi(k)$ is called the **morphological pattern spectrum** of image A and is used in texture classification schemes.

Two often-used suitable choices for the sequence $\{B_k\}$ are vertical and horizontal line segments of increasing length. The resulting opening sequences are called **linear granulometries.** Figure 5.28 shows two granular images, the one in part (a) exhibiting more horizontal length than the one in part (b). Figure 5.29 shows the corresponding horizontal, linear pattern spectra derived from each image of Fig. 5.28. Notice how the greater horizontal length of the image in Fig. 5.28(a) has resulted in a pattern spectrum whose main growth is to the right of the main growth region for the pattern spectrum corresponding to the image in Fig. 5.28(b).

Ignoring initialization and letting $B[k]$ denote the horizontal pixel line of length k, a direct implementation of the definition leads to the following algorithm to compute a horizontal linear size distribution:

```
while omega[k] > 0
    begin
      A[k + 1] := open(A,B[k+1])
      omega[k+1] := count(A[k+1])
    end
```

where `open` is a procedure to perform morphological opening and `count` is a procedure to count the number of points. The `while` loop is bounded by the horizontal dimension of the image frame; nevertheless, whether each opening is computed via the fitting expression of Eq. 5.7 or as an erosion followed by dilation, the algorithm is computationally burdensome on standard SISD hardware.

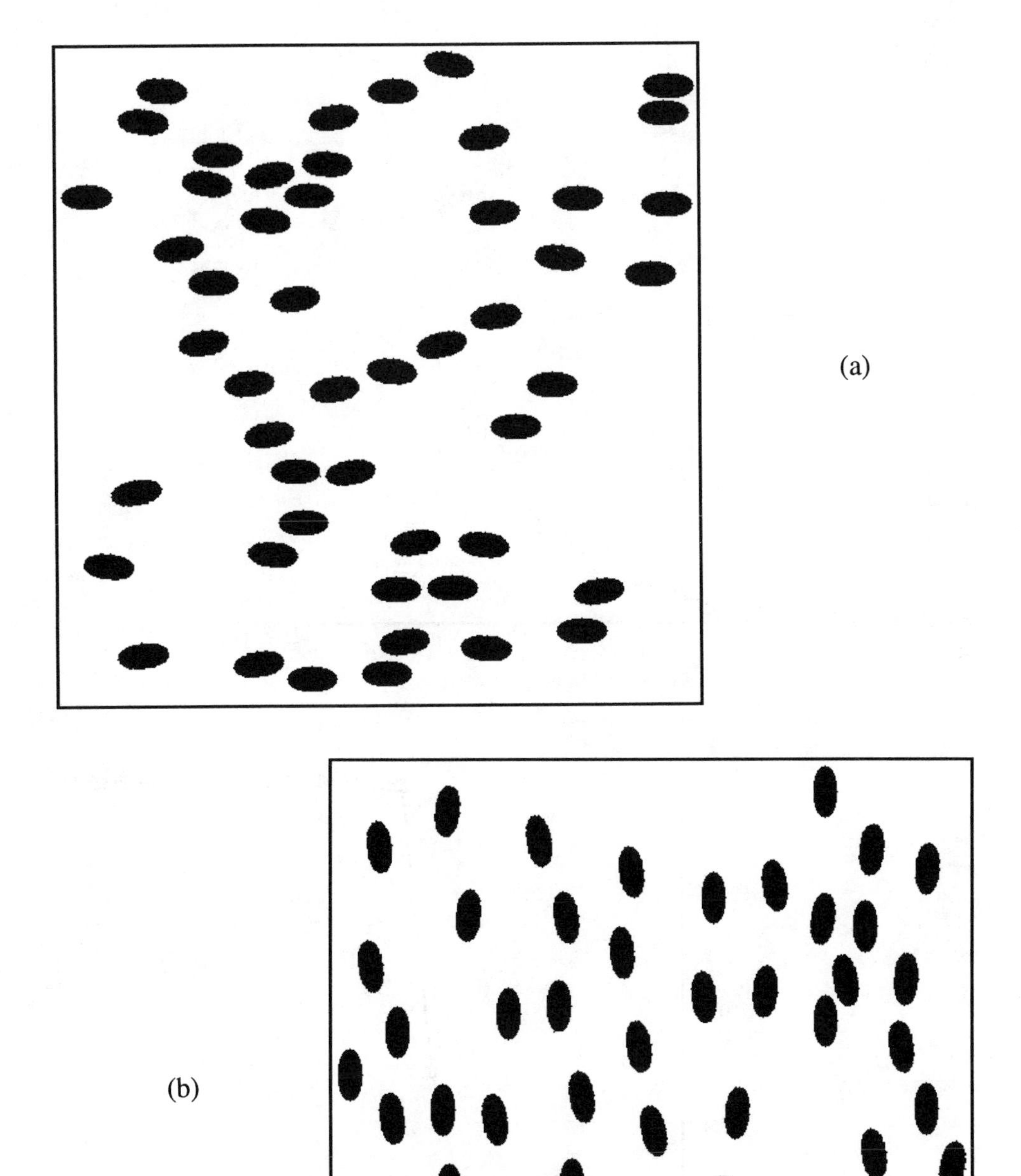

Figure 5.28: Two granular images: (a) exhibiting more horizontal length; (b) less horizontal length.

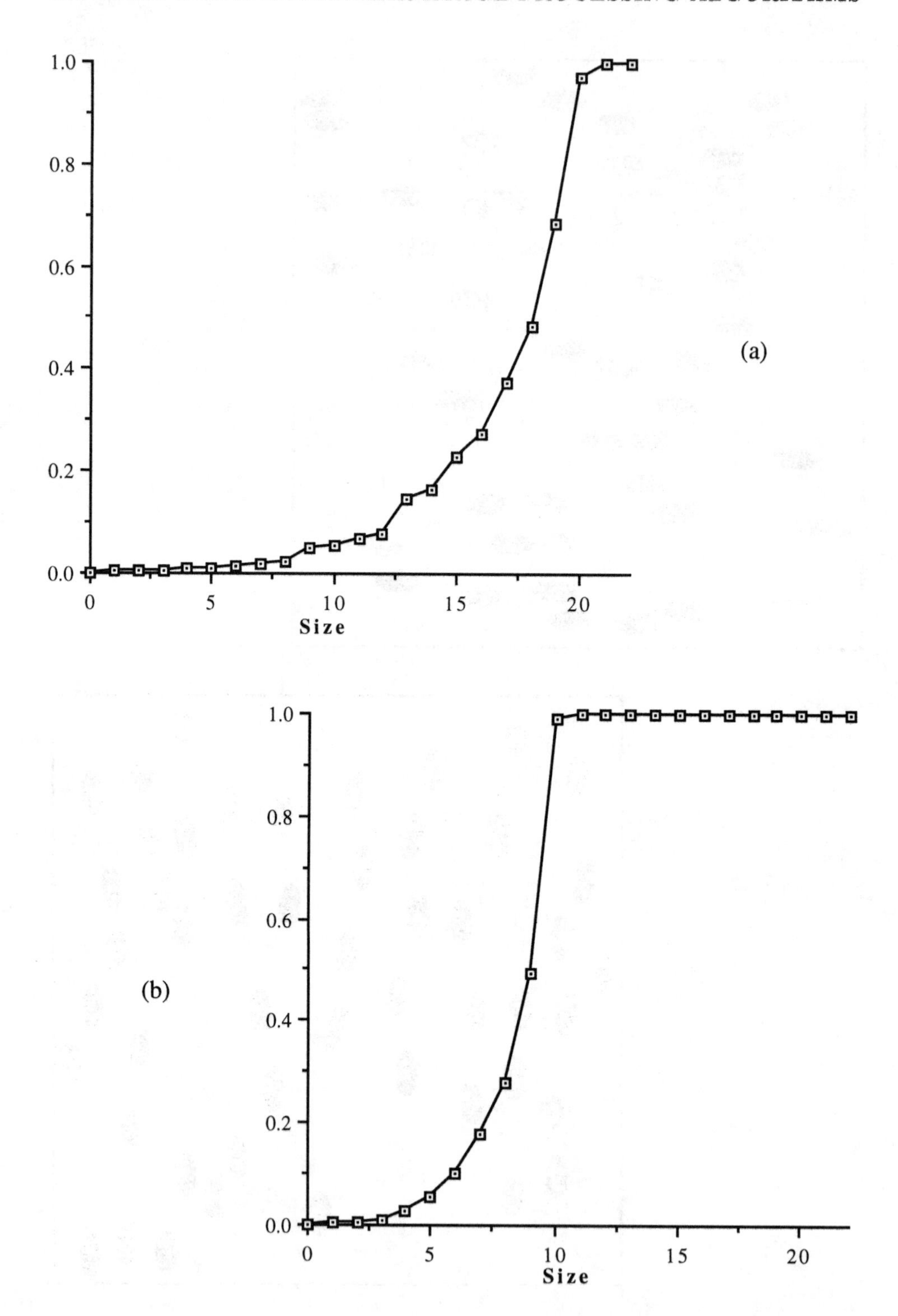

Figure 5.29: Pattern spectra for two granular images: (a) for Fig. 5.28(a); (b) for Fig. 5.28(b).

Suppose the binary image A is in a **run-length encoded** form. Specifically, the image is stored as a sequence of triples($x[i], y[i], r[i]$),$i = 1, 2, \ldots, n$, where $(x[i], y[i])$ specifies a pixel at which the image is black such that the pixel to the left is white, and $r[i]$ specifies the run length of the black pixels emanating at $(x[i], y[i])$. Such a compression scheme is lossless because the exact image can be reconstructed from the run-length encoding. The following algorithm can be employed to obtain the size distribution directly from the run-length encoding without recourse to actual image openings:

```
max := maximum(r[1],r[2],...,r[n])
for k = 1 to max
 for i = 1 to n
   if r[i] = k then
       temp[k] := temp[k] + k
   omega[k+1] := omega[k] - temp[k]
```

Here `maximum` is a procedure or macro to find the maximum of all the run-lengths. This improved code is significantly faster because of the aforementioned computational complexity of performing opening either using set operations or even the Boolean logic equivalent.

Chapter 6

Parallel Architectures

In this chapter we reexamine advanced architectures such as the MISD and MIMD machines introduced in Chapter 2 and show how they can be employed in conjunction with the image processing algorithms previously introduced to achieve enhanced real-time performance.

6.1 Pipelining

In Chapter 2 we defined pipelining as a type of MISD processing. Intuitively, then, we should be able to take advantage of this inherent parallelism. To illustrate how pipelining can enhance real-time performance, we consider the three-point convolution of a digital signal $x_1, x_2, x_3, \ldots$ by the mask with weights a_1, a_2, and a_3. A suitable pipeline architecture is given in Fig. 6.1, where MULT(a_i) means multiply the input from the register by a_i. At each clock pulse, two successive signal values are read into registers R1 and R2. From the fourth clock pulse onwards, the desired convolution value

$$C_k = a_1 x_k + a_2 x_{k+1} + a_3 x_{k+2} \tag{6.1}$$

is in R8 and is sent onto the bus at the next pulse. Figure 6.2 shows the contents of each register after each clock pulse.

Pipelining is an implicit parallelization of a serial instruction stream and, in general, improves average case performance by reducing the average instruction execution time. However, there are a couple of considerations. First, as discussed in Chapter 2, when doing any form of execution time analysis, pipeline flushing must be accounted for. Finally, there is a small

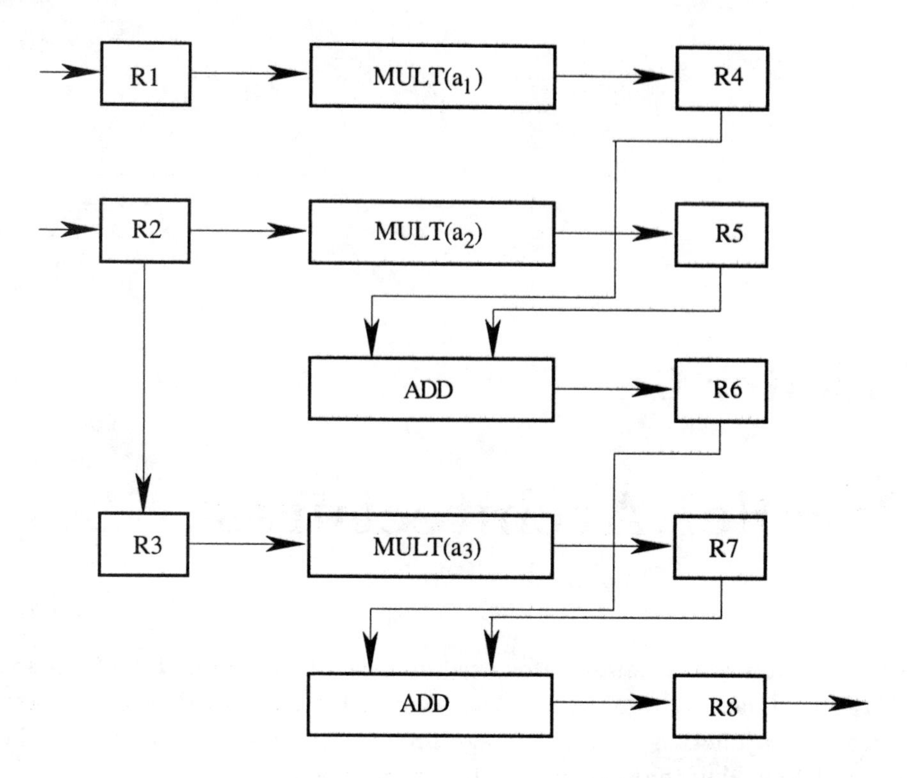

Figure 6.1: Convolution pipeline.

	$R1$	$R2$	$R3$	$R4$	$R5$	$R6$	$R7$	$R8$
1	x_1	x_2						
2	x_2	x_3	x_3	a_1x_1.	a_2x_2			
3	x_3	x_4	x_4	a_1x_2	a_2x_3	$a_1x_1 + a_2x_2$	a_3x_3	
4	x_4	x_5	x_5	a_1x_3	a_2x_4	$a_1x_2 + a_2x_3$	a_3x_4	c_1
5	x_5	x_6	x_6	a_1x_4	a_2x_5	$a_1x_3 + a_2x_4$	a_3x_5	c_2
6	x_6	x_7	x_7	a_1x_5	a_2x_6	$a_1x_4 + a_2x_5$	a_3x_6	c_3

Figure 6.2: Register content after each clock pulse.

penalty to pay in setup times, so that if the instruction stream were to execute in serial and not take advantage of the pipeline, the execution time would be longer than if the pipeline were not present at all. However, this is a pathological condition that is of little concern.

6.2 Dataflow Systems

Dataflow architectures can be used to enhance the real-time performance of inherently parallel algorithms. For example, in Section 3.3, we discussed edge detection via the Sobel and Prewitt gradients. To illustrate dataflow architecture we shall use it to implement the **Roberts gradient**, which employs Eq. 3.5 and the masks

$$g_1 = \begin{bmatrix} -1 & 0 \\ 0 & \mathbf{1} \end{bmatrix} \quad g_2 = \begin{bmatrix} 0 & 1 \\ -1 & \mathbf{0} \end{bmatrix} . \tag{6.2}$$

The output of the gradient is formed according to Eq. 3.5. Geometrically, it computes differences along the 45° lines, rather than along the horizontal and vertical directions. Computation of the gradient at a pixel (i, j) involves four values of the image f. For g_1, the gradient employs $f(i, j)$ and $f(i - 1, j + 1)$; for g_2, it employs $f(i, j + 1)$ and $f(i - 1, j)$.

A dataflow program for computation of the Roberts gradient at a pixel (i, j) is provided in Fig. 6.3. The opcodes have the following interpretations:

`TRANSLATE`: Evaluate f at the pixel obtained from (i, j) by translating by input values.

`MINUS`: Negate input value.

`ADD`: Add input values.

`MAX`: Take maximum of input values.

6.3 Systolic Array Processors

Because of the inherent parallel nature of many image processing algorithms, systolic architectures are ideal candidates for real-time implementations. The present section provides systolic architectures for multiplication and image convolution.

For matrix multiplication (which is used frequently in image processing such as in compression using Hadamard matrices), each cell is of the kind

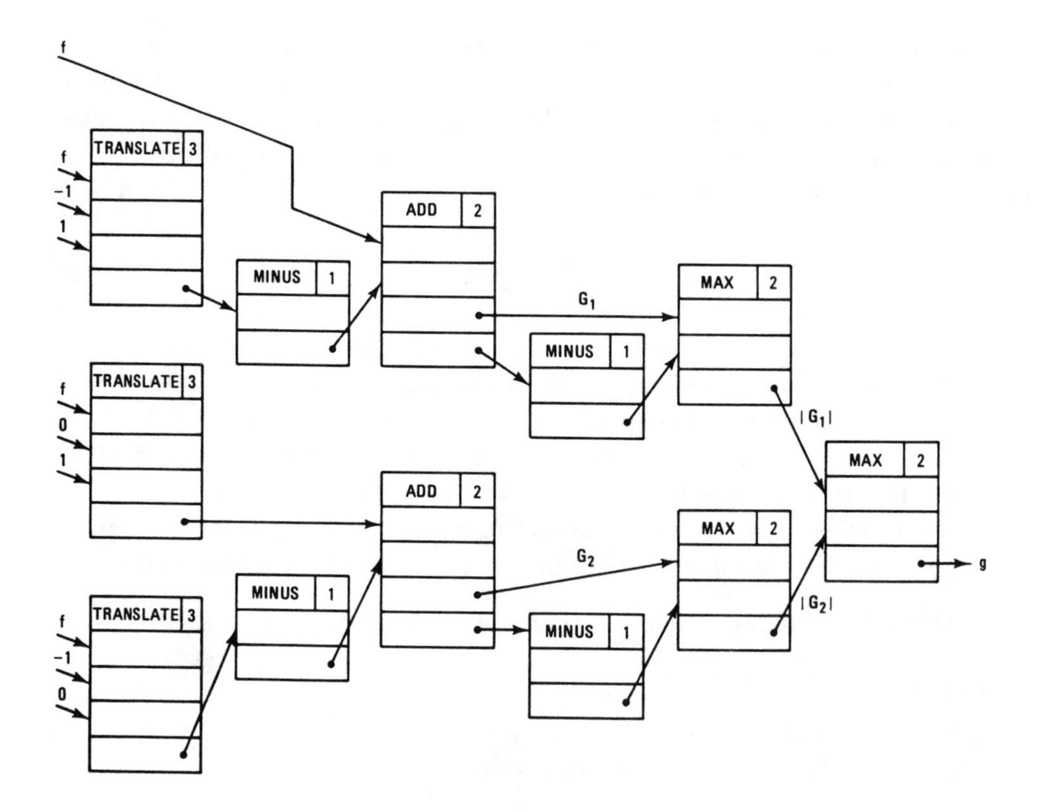

Figure 6.3: Dataflow implementation of Roberts gradient.

shown in part (a) of Fig. 6.4. It has a left input y_i and an upper input x_i. Outputs appear to the right (z) and at the bottom (u). The lower output equals the upper input unchanged, but delayed by one pulse. The right output equals the sum of the left input with the product of c and the upper input, so that $z = cx_i + y_i$. The outputs after one and two clock pulses are shown in parts (b) and (c) of Fig. 6.4.

Cells are connected to perform various specialized operations. Usually the right output of a cell is connected to the left input of another cell and the lower output is connected to the upper input. Figure 6.5 illustrates horizontal connection.

Consider the product of two matrices:

$$\begin{bmatrix} a_{11} & a_{12} \\ a_{21} & a_{22} \end{bmatrix} \times \begin{bmatrix} b_{11} & b_{12} \\ b_{21} & b_{22} \end{bmatrix} = \begin{bmatrix} c_{11} & c_{12} \\ c_{21} & c_{22} \end{bmatrix}. \tag{6.3}$$

The output of a systolic array is determined by the cell function, the connections, and the initial configuration. For the matrix multiplication of Eq. 6.3, the initial configuration is shown in Fig. 6.6 and the configurations and upper and lower channel outputs after each of four clock pulses are shown in Figs. 6.7 through 6.10.

Next we consider a systolic-array implementation for image convolution. Although the general scheme applies to masks of any size, for the sake of clarity and space, we show the architecture for the case of a 3×3 mask. The basic cell functions are the same as for matrix multiplication (Fig. 6.4) except that there is no lower output. Both image and mask are represented as matrices. When the mask center is situated over a pixel adjacent to the image matrix, the mask continues to intersect the image so that the boundary effect is that the image frame is expanded. The systolic array will output the expanded image; however, if desired, the extra two rows and columns can later be deleted. Given the operation of the systolic cells, the array is defined by its geometry and initialization. Input timing is employed in conjunction with array geometry to make certain the array outputs values consistent with mask positioning as it moves across the image. The position of the image in the grid is irrelevant to the computational scheme.

Let the input image f be defined by the m by n matrix

$$f = \begin{bmatrix} a_{1,1} & a_{1,2} & \cdots & a_{1,n} \\ a_{2,1} & a_{2,2} & \cdots & a_{2,n} \\ \vdots & \vdots & & \vdots \\ a_{m,1} & a_{m,2} & \cdots & a_{m,n} \end{bmatrix} \tag{6.4}$$

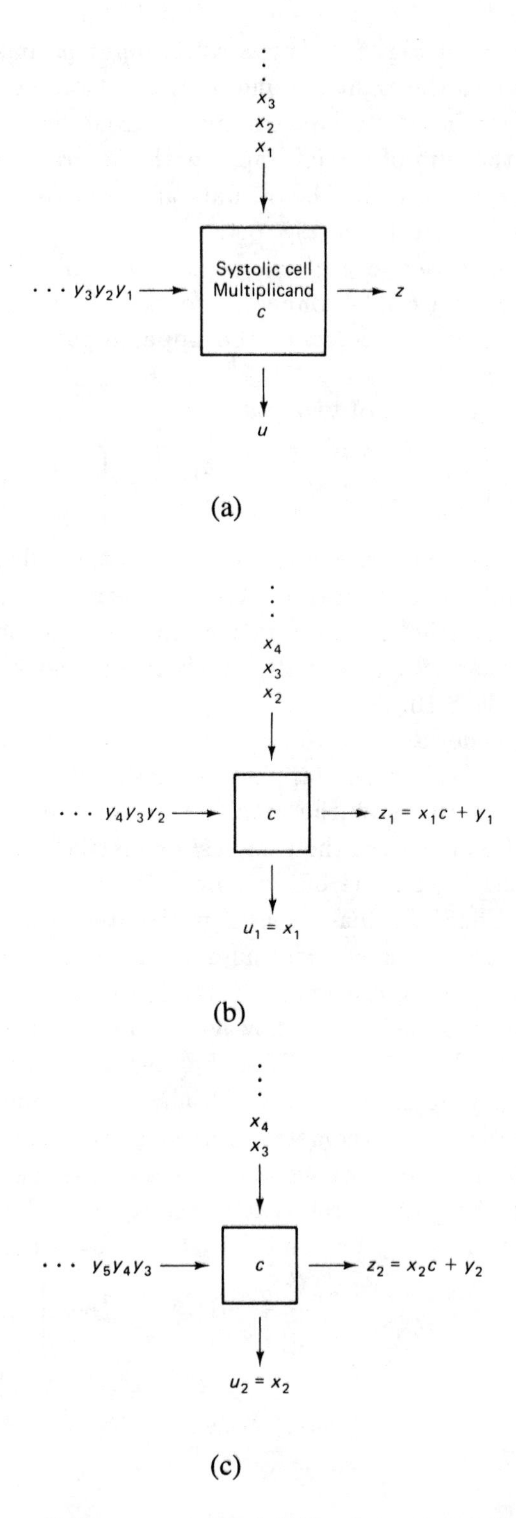

Figure 6.4: Systolic cell for matrix multiplication: (a) cell structure; (b) output after first clock pulse; (c) output after second clock pulse.

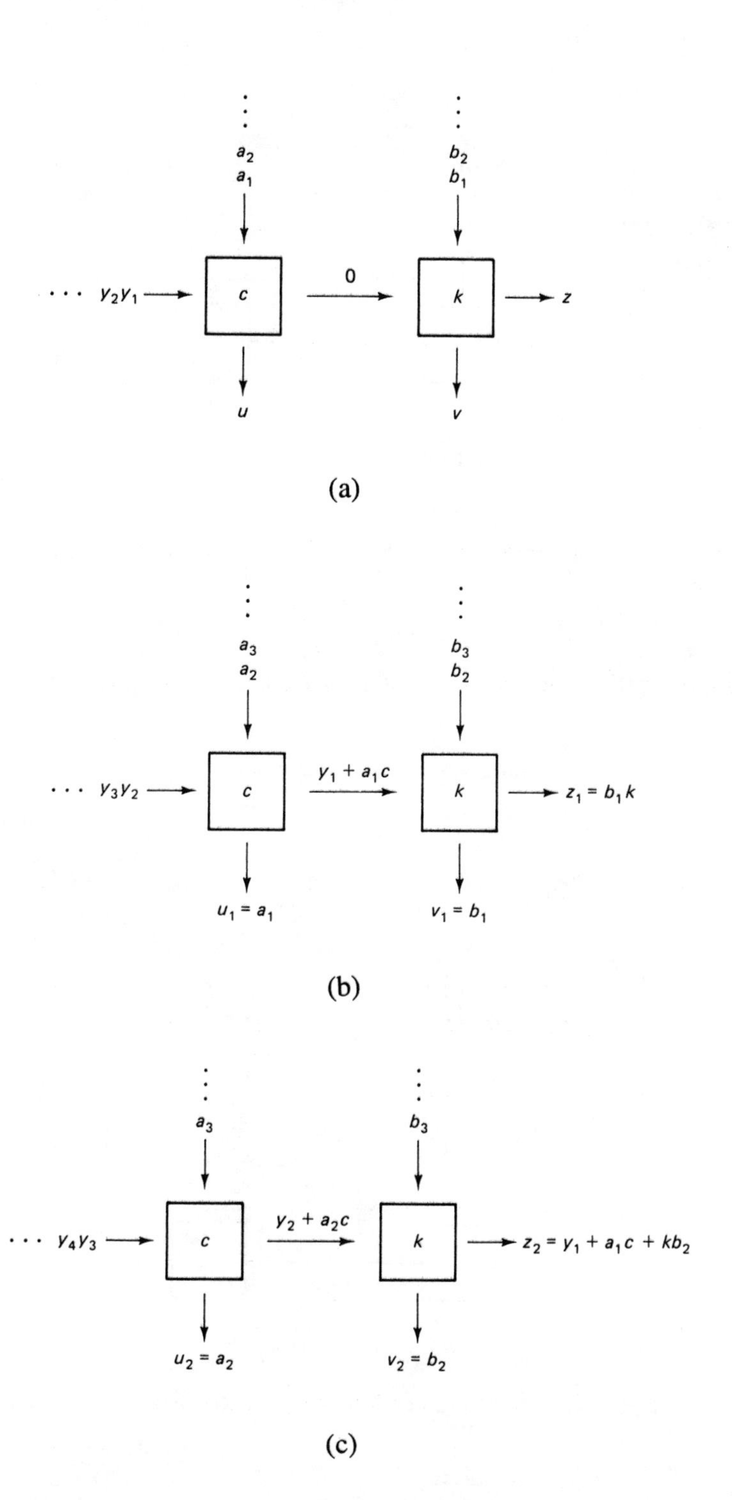

Figure 6.5: Horizontal cell connection: (a) initial configuration; (b) after first clock pulse; (c) after second clock pulse.

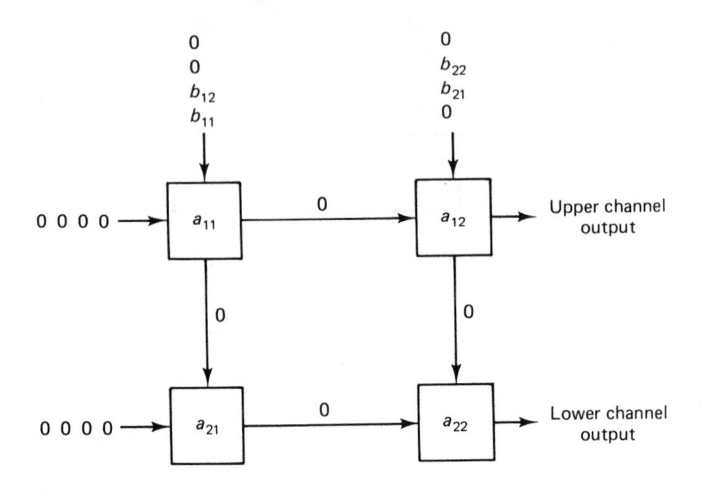

Figure 6.6: Initial configuration of systolic array for matrix multiplication.

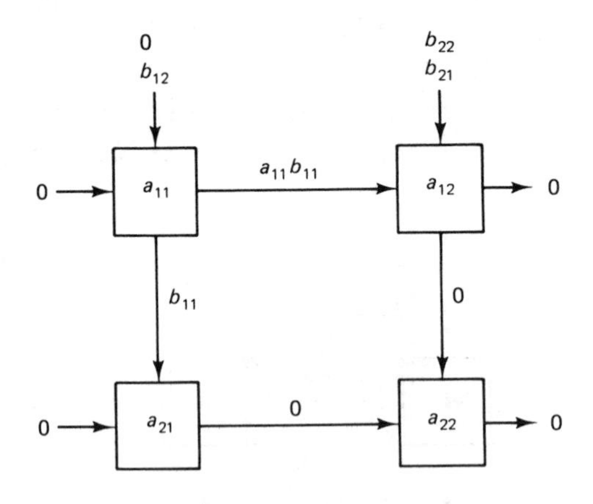

Figure 6.7: Systolic configuration after first clock pulse.

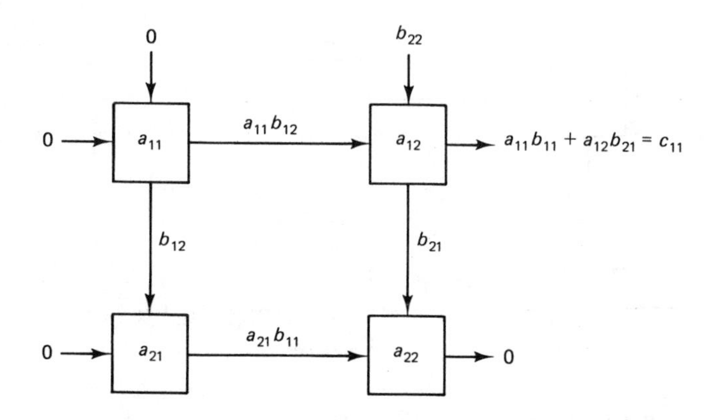

Figure 6.8: Systolic configuration after second clock pulse.

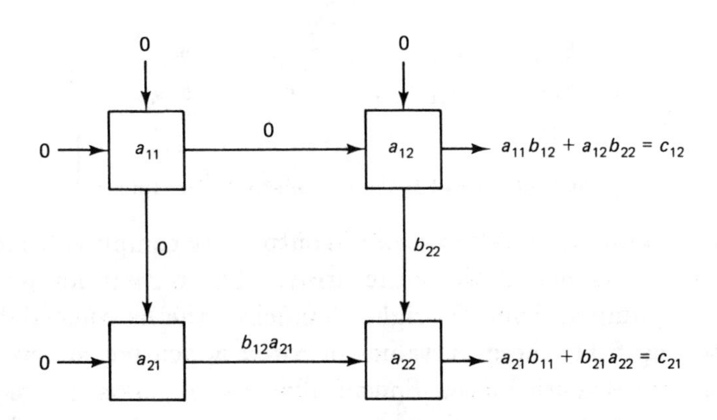

Figure 6.9: Systolic configuration after third clock pulse.

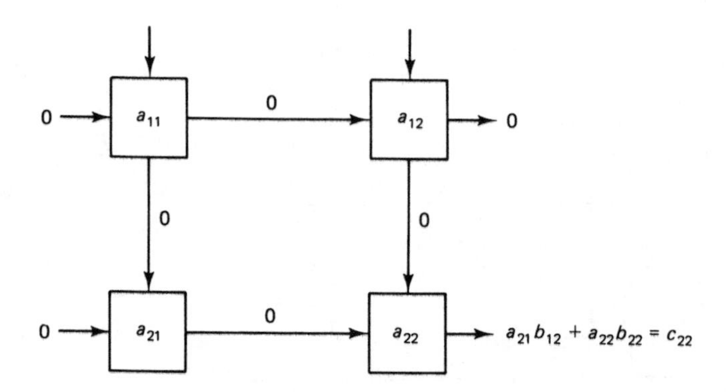

Figure 6.10: Systolic configuration after fourth clock pulse.

and the mask be defined by

$$g = \begin{bmatrix} c_{11} & c_{12} & c_{13} \\ c_{21} & \mathbf{c_{22}} & c_{23} \\ c_{31} & c_{32} & c_{33} \end{bmatrix}. \tag{6.5}$$

Because g is 3×3, the output is $(m+2) \times (n+2)$:

$$h = \begin{bmatrix} b_{0,0} & b_{0,1} & \cdots & b_{0,n} & b_{0,n+1} \\ b_{1,0} & b_{1,1} & \cdots & b_{1,n} & b_{1,n+1} \\ \vdots & \vdots & & \vdots & \vdots \\ b_{m+1,0} & b_{m+1,1} & \cdots & b_{m+1,n} & b_{m+1,n+1} \end{bmatrix}. \tag{6.6}$$

In the execution, the $m+2$ rows of the output are computed concurrently by having $m+2$ copies of the basic array. The output for row i, $0 \leq i \leq m+1$, is pumped from the right channel of the extreme right cell in the array of Fig. 6.11, the gray values of row i appearing in reverse order beginning at the seventh pulse. Specifically, $n+8$ pulses are required to complete the computation of the convolution. Given the operation of the cells, processing accomplished by the array is determined by the initialization given in Fig. 6.11, where each upper input channel is fed concurrently into the

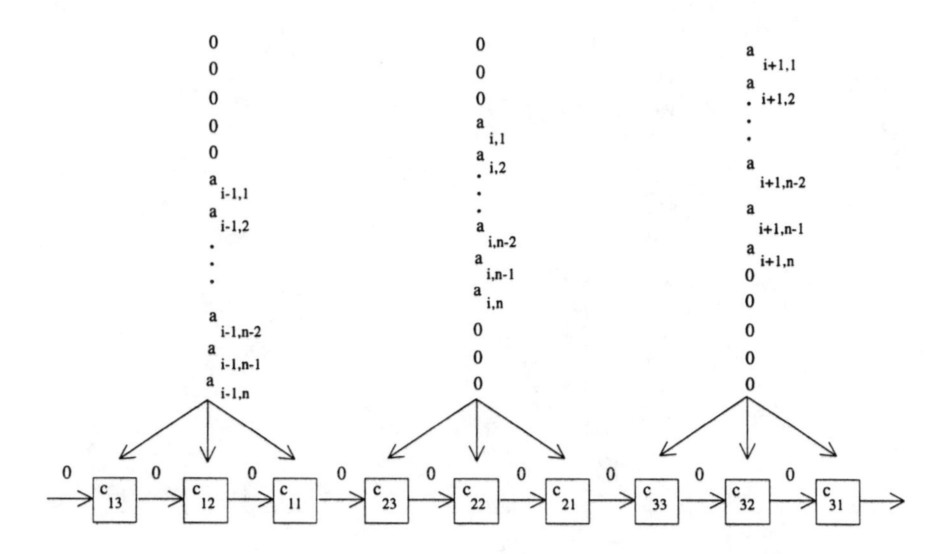

Figure 6.11: Systolic array initialization for 3 × 3 convolution.

block of three cells below it. Note that each left input channel is initialized to 0 and 0 has also been employed to stagger the inputs so that proper timing is affected. Because the output matrix is larger than the input matrix and we need to consider the rows $i-1$ and $i+1$, it is necessary to augment the matrix of f with two rows of 0s on top and bottom for the cases in which the top and bottom two rows of h are being computed. For instance, the 0th row of f consists of all 0s. Given these conventions, the initialization of Fig. 6.11 defines the systolic array for convolution by a 3 by 3 mask.

Based on the linear array of Fig. 6.11, the output for row i is

$$0\ 0\ 0\ 0\ 0\ 0\ b_{i,n+1}\ b_{i,n}\ \ldots\ b_{i,0}.$$

This output is a string of $n+8$ characters, the last $n+2$ of which define the ith row of the output image. For instance, if the original image is 256 by 256 and the rows are executed concurrently (this means that the hardware would consist of 258 replications of Fig. 6.11), then the output matrix is 258 by 258 and 264 pulses are required to execute the convolution.

As noted previously, the scheme applies to other sized masks. For in-

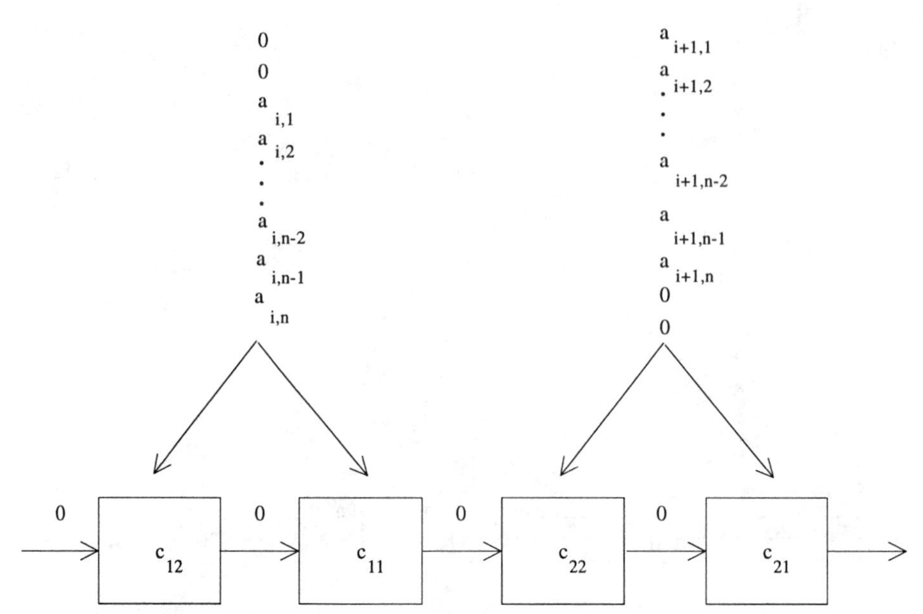

Figure 6.12: Systolic array initialization for 2×2 convolution.

stance, for the 2 by 2 mask

$$g = \begin{bmatrix} c_{11} & c_{12} \\ c_{21} & c_{22} \end{bmatrix} \tag{6.7}$$

the output takes the form

$$h = \begin{bmatrix} b_{1,0} & b_{1,1} & \dots & b_{1,n} \\ \vdots & \vdots & & \vdots \\ b_{m,0} & b_{m,1} & \dots & b_{m,n} \\ b_{m+1,0} & b_{m+1,1} & \dots & b_{m+1,n} \end{bmatrix}. \tag{6.8}$$

The initialized array is shown in Fig. 6.12. Row i output is

$$0\ 0\ b_{i,n}\ b_{i,n-1} \dots\ b_{i,0}.$$

The computation requires $n + 3$ clock pulses.

We illustrate the method for image and mask

$$f = \begin{bmatrix} 3 & 1 & 2 \\ 1 & 2 & 1 \end{bmatrix} \quad g = \begin{bmatrix} 1 & 2 \\ 3 & 4 \end{bmatrix} \tag{6.9}$$

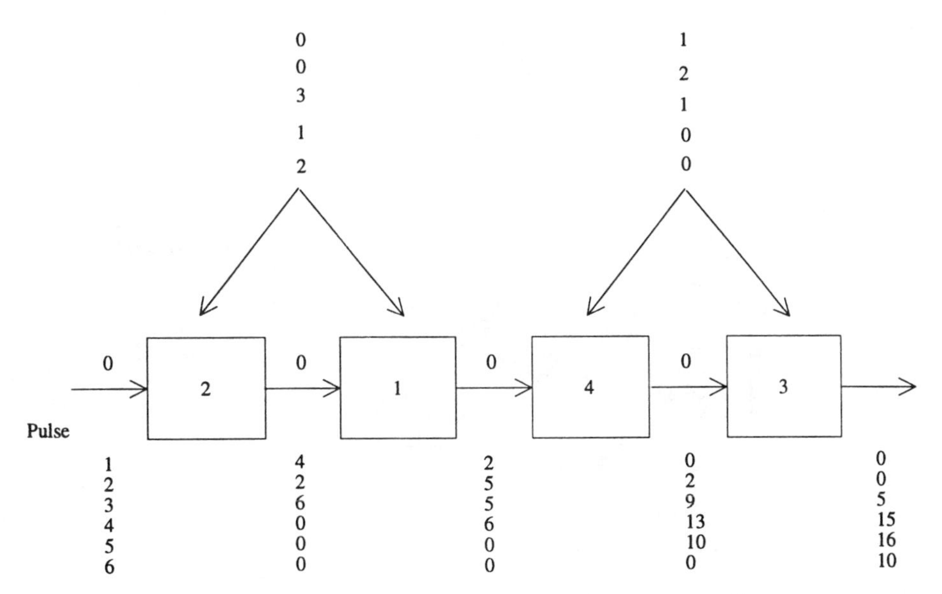

Figure 6.13: Example of 2 × 2 convolution systolic array.

for which the convolved image is

$$f * g = \begin{bmatrix} 12 & 13 & 11 & 6 \\ 10 & 16 & 15 & 5 \\ 2 & 5 & 4 & 1 \end{bmatrix}. \tag{6.10}$$

Application of the array of Fig. 6.12 to obtain the second row of $f * g$ is illustrated in Fig. 6.13. Note that the last pulse is employed to empty the pipeline.

Although we shall not discuss the matter in the present text, there exist gray-scale erosion and dilation operations besides the flat erosion and dilation defined in Section 5.7 via moving minimum and maximum. These "true gray-scale" morphological operators have structuring elements that are gray-scale templates in much the same way as convolution templates. Both gray-scale dilation and erosion possess systolic implementations that are very similar to systolic implementation of convolution. Flat erosion and flat dilation systolic implementations are special cases of the general systolic morphological implementations.

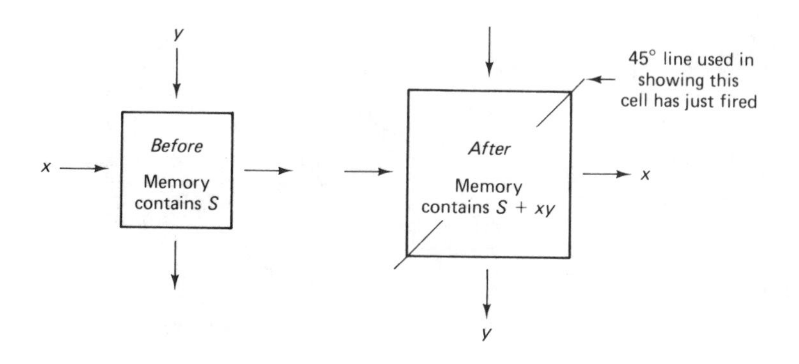

Figure 6.14: Wavefront cell for matrix multiplication.

6.4 Wavefront Array Processors

Again, because it is highly parallel, a wavefront processor can often be used to improve real-time imaging performance. We illustrate a wavefront architecture for the matrix multiplication of Eq. 6.3.

Each cell is programmed to perform the same simple operation and has left and upper input channels. When (and only when) both inputs are present, the processor multiplies the inputs together and adds the result to the contents of an internal accumulator within the cell. The left and upper inputs are sent out the right and lower channels, respectively. Cell operation is illustrated in Fig. 6.14. A massively parallel special-purpose computer is formed by connecting numerous cells in an appropriate array geometry.

Figure 6.15 shows the wavefront array initialization for the matrix multiplication of Eq. 6.3 and Figs. 6.16 through 6.19 show the states of the array following firings of the upper left cell, both left cells and the upper right cell, the lower left cell and both right cells, and the bottom right cell, respectively. The desired result for the wavefront array appears inside each cell after all the inputs have been used up, that is, after each cell has executed twice. Cells execute at different times from one another: there is no external clock. At the outset only the left upper cell can fire because only it has both

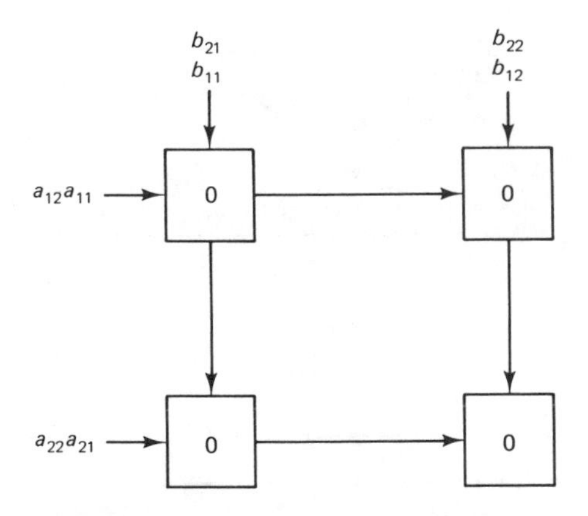

Figure 6.15: Initialization of wavefront array for matrix multiplication.

operands available. Then all cells except the bottom right can fire because all but it have both operands present. This is shown in Fig. 6.17 where there are two computation waves going across the array. Next, all cells but the upper left can fire, and finally only the bottom right cell can fire, thereby completing the computation.

6.5 Linear Array, Mesh, and Hypercube Processors

Some commercial real-time imaging processors employ a one-dimensional array of simple processing elements (PEs) where the size of the array is as wide as the image. These fine-grained SIMD architectures, called **linear array processors** or **vector processors**, have one PE for each column in the image, and enough memory directly connected to each PE to hold the entire column of image data for several distinct images (see Fig. 6.20). A single image operation is processed by first broadcasting an instruction to all image PEs. The image data are sent row-by-row to the processing elements, and the results are returned row-by-row back to the memory. Nearest neighbor information is exchanged between adjacent processors during the processing.

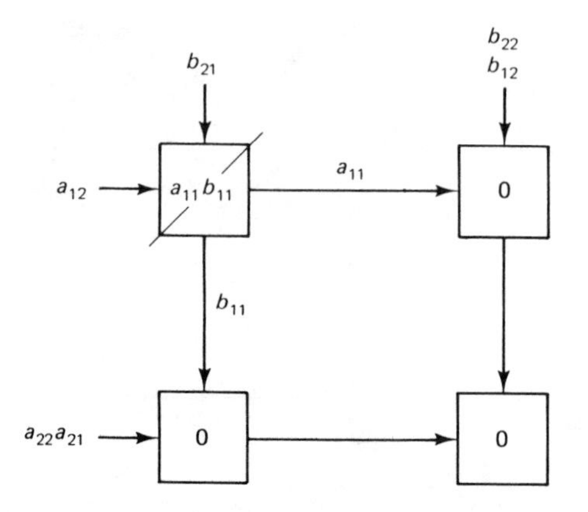

Figure 6.16: Array configuration after upper left cell has fired.

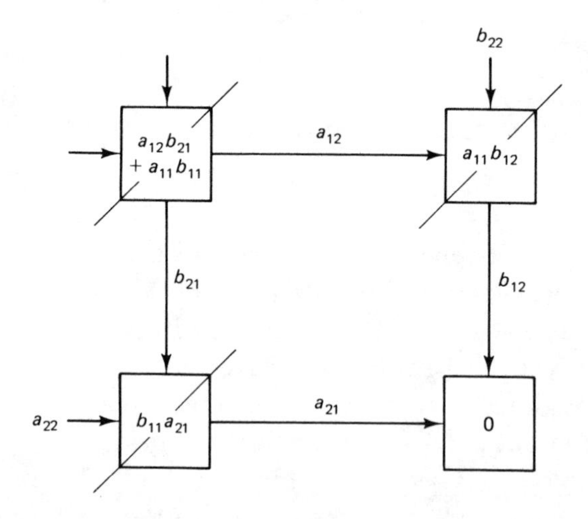

Figure 6.17: Array configuration after upper right and left cells have fired.

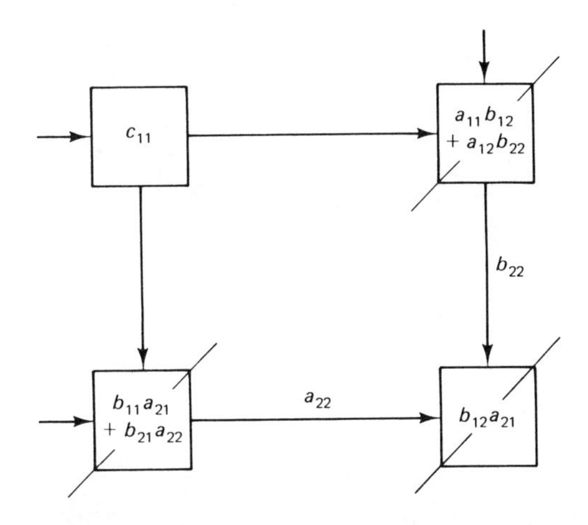

Figure 6.18: Array configuration after lower left and right cells have fired.

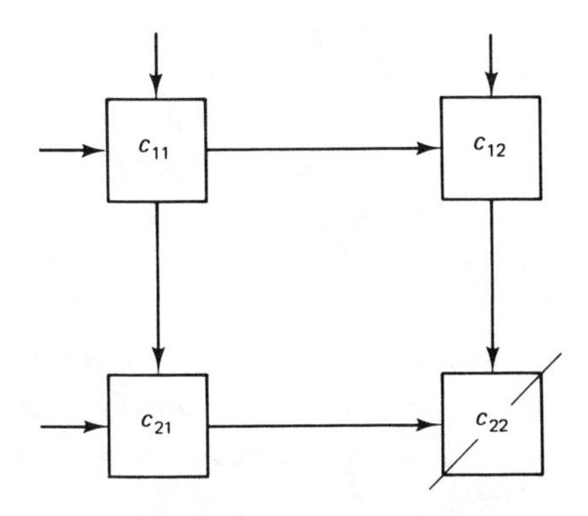

Figure 6.19: Array configuration after lower right cell has fired.

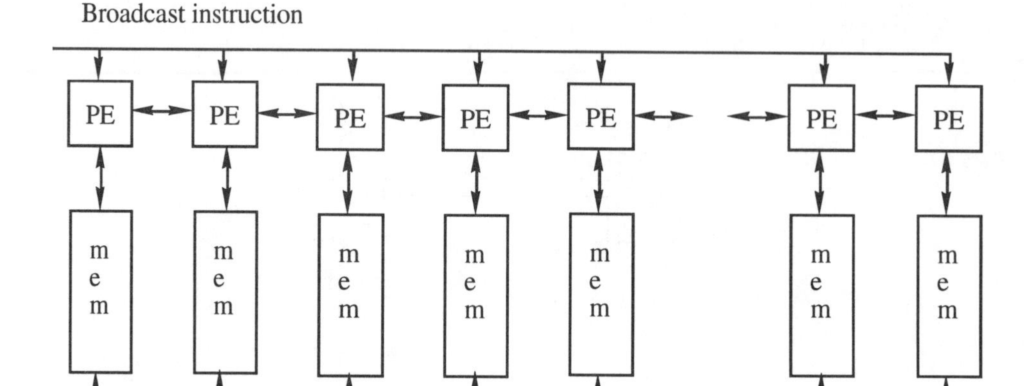

Figure 6.20: Linear array architecture.

One advantage of such an arrangement is that complex instructions can be executed at a rapid pace since they are applied in parallel to all pixels in a row. A regular pattern of memory addresses is all that is needed at high execution rates. In addition, the linear structure is a natural architecture for scanned data as it comes from a camera.

A notable class of a commercial fine-grained SIMD architectures is the Applied Intelligent Systems, Inc. (AISI) linear array processors. These processors employ a linear array of PEs, where the size of the array is 64 (AIS-3000), 128 (AIS-3500), and 512 (AIS-4000). In the AIS-4000 there are enough PEs so that the array is as wide as the image. The image storage is such that, generally, 128 binary images, 16 gray-scale images, or any combination of various image word sizes can be saved. A general-purpose host microcomputer controls the sequence of the image processing functions, and can also read or write data directly to the image memory. A controller is connected to a small memory that contains microcode instructions and general data for support of the parallel processing array. As a hardware background task, a camera may input data to a separate buffer in the host computer. When an entire line of data has been captured, the processor is interrupted and these data are sent to the image memory. C language software allows for flexible programming support.

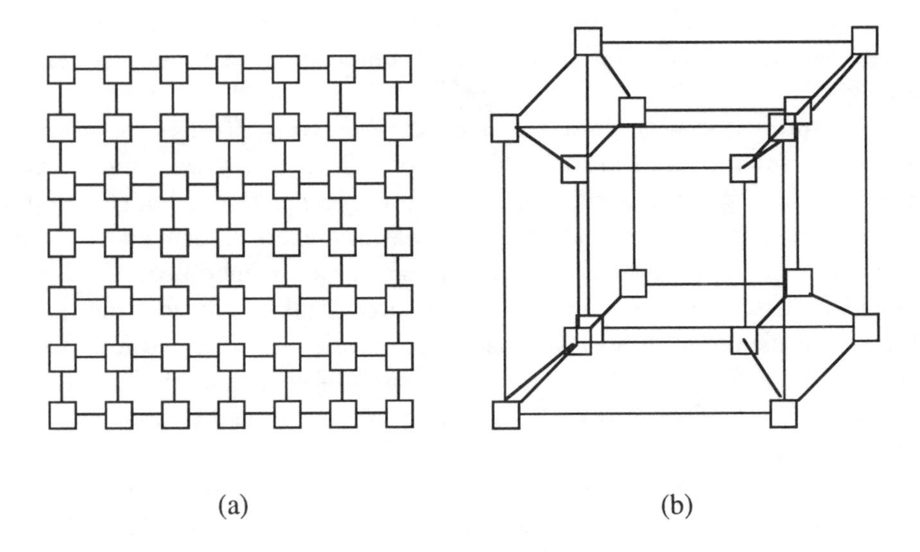

Figure 6.21: Interconnection structures: (a) mesh; (b) hypercube.

Two other coarse-grained SIMD architectures are the **mesh processor** and **hypercube processor**. The mesh and hypercube configurations are similar to the linear array processor except that they have a much higher band pass in data communications. Each linear array PE communicates data only to nearest-neighbors to the east and west. The mesh system also communicates data north and south. A hypercube communicates data along a number of other higher dimensional pathways (see Fig. 6.21). In principle, but not in practice, the mesh and hypercube systems can grow without bounds until there is one PE per pixel. The chief disadvantages of these architectures are that complex controllers are needed to update the instruction stream at high speed, the instructions must be kept simple due to the limited number of pins on the chip, complex hardware support is needed to load an image into the arrays, and images are always larger than the array size.

6.6 Associative Memory

For many CPU instructions, memory access times represent a bottleneck. Hence, higher performance can be achieved through enhanced memory access techniques. The present section discusses one such scheme, **associative memory** (also called **content-addressable memory**), which employs a storage device that achieves multiple-data processing via a built-in search capacity.

When presented with data, an associative-memory processor can automatically compare that data with data stored in memory and mark those locations where there is agreement. Moreover, associative-memory processors can write in parallel into all memory locations where agreement has been determined, and part or all of the agreeing words can be modified. Hardware demand for associative memory is only slightly greater than for coordinate addressable memory. Associative memories work well where comparisons can be made on the basis of a subset of a word's bits (such as the high-order bits) as in thresholding, stack filters, fitting in erosion, and implementation of minimum and maximum.

Besides the location where information is stored, there are essentially three other components to the associative memory structure (Fig. 6.22). There is the comparand, where information to be retrieved is placed, and a **mask register**, which is used to block out certain bits in the comparand that are not to be checked in the retrieval process. When a `COMPARE` instruction is issued, certain cells in memory are marked, these being the cells containing a word that matches the unmasked bits in the comparand. Memory cells are marked using a tag memory, known as a **response store**. Those places where a match exists are set equal to 1 in the response store.

We illustrate the manner in which an associative-memory processor can be used to compute a minimum, which is a key operation in nonlinear image processing. Each word of memory consists of n bits, is binary, and is interpreted as a nonnegative integer. The algorithm involves n concurrent searches of memory, regardless of the number of data entries.

Initially all tag bits are set to 1, the comparand is set to all 0s, and the mask register has a 1 in the most significant bit and 0s elsewhere. A search is conducted. If any tag bits remain set to 1, then there has been a match and there must exist entries possessing a 0 in the most significant bit. Thus, the most significant bit of the minimum in the database is 0. If no match is obtained, then it is known that the most significant bit in the minimum is 1 and the comparand is changed to reflect this conclusion. In either event, the highest address bit of the minimum has been found. The next highest address bit is found next: the mask is altered so as to have two 1s in the two highest-ordered bits, and a second search is performed. The second highest bit in the comparand is then adjusted to generate a match. Recursively, if

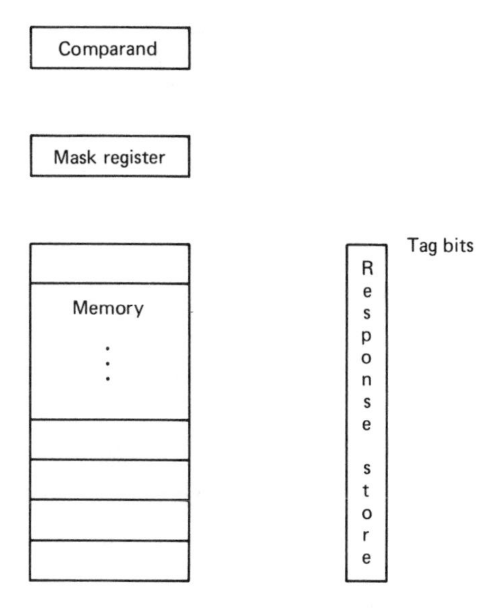

Figure 6.22: Structure of associative memory.

a match is found, then change the leftmost 0 in the mask to 1; if a match is not found, then change the 0 in the comparand above the rightmost 1 in the mask to 1 and change the leftmost 0 in the mask to 1. Do this n times. The minimum value in memory will be contained in the comparand at the conclusion of the nth step.

A walk-through of the minimum procedure is given in Fig. 6.23. Part (a) shows the contents of memory and part (b) shows the comparand and mask at initialization and subsequent to each of the four searches. Using the initial settings, there is a match, so the second most significant bit of the mask is set to 1; using the resulting settings, there is a match, so the third most significant bit of the mask is set to 1; using the resulting settings, there is no match, so the third most significant bit of the comparand is set to 1 and the remaining bit of the mask is set to 1; using these settings, the last search yields a match, so the last bit of the comparand is set to 1 and the comparand contains the minimum $0011 (= 3)$.

15	1	1	1	1
3	0	0	1	1
12	1	1	0	0
8	1	0	0	0
5	0	1	0	1
9	1	0	0	1

MEM

(a)

Initial	Comparand	0 0 0 0
	Mask	1 0 0 0
1	Comparand	0 0 0 0
	Mask	1 1 0 0
2	Comparand	0 0 0 0
	Mask	1 1 1 0
3	Comparand	0 0 1 0
	Mask	1 1 1 1
Final(4)	Comparand	0 0 1 1
	Mask	1 1 1 1

(b)

Figure 6.23: Finding a minimum using associative memory.

Chapter 7

Programming Languages

In many settings, specialized parallel architectures used to enhance real-time performance are not available or cost-effective. In these cases, we are forced to use von Neumann processors and the standard programming languages that support them.

Until now we have been talking mostly about designing, coding, and optimizing image processing algorithms and applications in assembly language. Modern software engineering, however, leads us to use more sophisticated tools such as high-level languages. In this chapter we discuss the effects on real-time performance that are introduced by the high-level language itself. We also discuss ways to persuade compilers to produce better code. We do this in a language-independent fashion, although we discuss issues arising in some of the more commonly used programming languages.

There are several programming language features that are desirable for real-time system use, including:

- versatile parameter passing mechanisms
- dynamic memory allocation facilities
- strong typing
- exception handling
- modularity.

Each feature has an impact on real-time performance and we discuss them in some detail in the following sections.

7.1 Parameter Passing Techniques

There are several methods of parameter passing, including the use of parameter lists and global variables. While each of these techniques is used in some specific way, each has a different real-time performance impact.

There are two widely implemented parameter passing methods: call-by-value and call-by-reference. In **call-by-value** parameter passing, the value of the actual parameter in the subroutine or function call is copied into the procedure's formal parameter. Since the procedure manipulates the formal parameter, the actual parameter is not altered. This is useful when either a test is being performed or the output is a function of the input parameters. For example, in an edge detection algorithm, an image is passed to the procedure and some description of the location of the edges is returned, but the image itself need not be changed. When parameters are passed using call-by-value they are copied onto the stack at run-time, at considerable execution time cost. For example, large image arrays must be passed pixel-by-pixel.

In **call-by-reference** or **call-by-address** the address of the parameter is passed by the calling routine to the called procedure so that it can be altered there. Execution of a procedure using call-by-reference can take longer than one using call-by-value since in call-by-reference indirect mode instructions are needed for any calculations involving the variables passed. However, in the case of passing images between procedures it is more desirable to use call-by-reference since passing a pointer to an image is more efficient than passing the image pixel-wise. For example, in a filtration algorithm an image along with a filter element can be passed to it, and a filtered image returned, e.g.,

```
call filter(image,filter_element,output)
```

where `image`, `filter_element`, and `output` are all pointers to large images. This is superior to passing each individual pixel by value.

Using parameter lists is a well-known technique for promoting modular design because the interfaces between the modules are clearly defined. Clearly defined interfaces reduce the potential of untraceable corruption of data by procedures using global access. However, both parameter passing techniques are costly when the lists are long since interrupts are frequently disabled during parameter passing to preserve the time correlation of the data passed.

Global variables are variables that are within the scope of all modules of the software system. This usually means that references to these variables can be made in direct mode and thus are faster than references to variables passed via parameter lists. For example, in many image processing applications, global arrays are defined to represent images, hence costly parameter passing can be avoided.

Global variables introduce no timing problems but are dangerous because reference to them may be made by unauthorized modules, thus introducing subtle bugs. For this and other reasons, flagrant use of global variables is to be avoided. Use global passing only when timing warrants and clearly document it – otherwise use parameter lists.

The decision to use one method of parameter passing or the other represents a trade-off between accepted software engineering methodology and speed. The final system usually employs all parameter passing techniques in some combination reflecting the trade-offs of good software engineering technique to the realities of timing constraints. For example, often timing constraints force the use of global parameter passing in instances when parameter lists would have been preferred for clarity and maintainability.

7.2 Recursion

Many programming languages provide a mechanism called **recursion**, whereby a procedure can call itself. Recursion is widely used in image processing, for example, in implementation of the Hadamard matrices, fractal compression, segmentation via divide-and-conquer methods, quad trees, and thinning.

As a simple illustration of recursion, consider the recursive Quicksort algorithm. Quicksort sorts a list of numbers `A(first) ... A(last)` into ascending order, an operation that has applications in image processing (for example, in finding the median for median filtering). `Partition` is a procedure that divides a sublist of numbers so that all entries to the left of a key entry are less than or equal to the key and all entries to the right are greater than the key. The key entry can be anywhere in the sublist. In pseudocode, Quicksort can be given by:

```
procedure Quicksort
   if first < last then
       key := last + 1
       Partition(first,key)      % partition list
       Quicksort(first,key-1)    % sort left part of list
```

```
        Quicksort(key+1,last)    % sort right half of list
    endif
```

While recursion is elegant and is often called upon in imaging algorithms, its adverse impact on real-time performance cannot be overemphasized. Procedure calls require the allocation of storage on one or more stacks for the passing of parameters and for storage of local variables. The execution time needed for the allocation and deallocation, and for the storage of those parameters and local variables, can be ill-afforded. In addition, recursion necessitates the use of a large number of costly memory and register indirect instructions. Finally, the use of recursion often makes it impossible to determine the size of run-time memory requirements. Thus, iterative techniques such as loops must be used if performance prediction is crucial or in those languages that do not support recursion.

7.3 Dynamic Memory Allocation

The ability to dynamically allocate memory is important in the construction and maintenance of stacks needed by the real-time operating system. While dynamic allocation can be time consuming, it is usually necessary, especially when creating intermediate images needed in most imaging algorithms. Linked lists, trees, heaps and other dynamic data structures used in real-time imaging applications can benefit from the clarity and economy introduced by dynamic allocation. And in cases where just a pointer is used to pass a data structure, then the overhead for dynamic allocation can be quite reasonable. When writing imaging programs care should be taken to ensure that the compiler will pass pointers to large data structures and not the data structure itself.

Languages that do not allow dynamic allocation of memory require data structures of fixed size. While this may be faster, flexibility is sacrificed and memory requirements must be predetermined.

Languages such as C, Pascal, Ada, and Modula-2 have dynamic allocation facilities while most versions of Fortran, for example, do not.

7.4 Typing

Typed languages require that each variable and constant be of a specific type (e.g., "pixel," "Boolean," and "integer") and that each be declared as such before use. Languages that provide specialized types for imaging applications are rare. Generally, high-level languages provide integer and floating point types, along with Boolean. In some cases, **abstract data types** are supported. These allow a programmer to define his or her own type (such as "pixel") along with the requisite operation that can be applied to it. Use of abstract data types, however, will incur an execution time penalty as complicated internal representations are often needed to support the abstraction. Strongly typed languages prohibit the mixing of different types in operations and assignments, and thus force the programmer to be precise about the way data are to be handled. Precise typing can prevent corruption of data through unwanted or unnecessary type conversion. Hence, strongly typed languages are desirable for real-time imaging.

Some languages are typed, but do not prohibit mixing of types in arithmetic operations. Since these languages generally perform mixed calculations using the type that has the highest storage complexity, they must promote all variables to that type. For example, in C, the following code fragment illustrates automatic promotion and demotion of variable types:

```
int i,j;
float k,l,m;
   .
   .
j = i*k+m
```

Here the variable `i` will be promoted to a `float` (real) type and then multiplication and addition will take place in floating-point. Afterward, the result will be truncated and stored in `j`. The real-time impact is that hidden promotion and more time-consuming arithmetic instructions can be generated, with no additional accuracy. In addition, accuracy can be lost due to the truncation, or worse, an integer overflow can occur if the floating-point value is larger than the allowable integer value. Programs written in languages that are weakly typed need to be scrutinized for such effects.*

7.5 Exception Handling

Certain languages provide facilities for dealing with errors or other anomalous conditions that arise during program execution. These conditions in-

*Some C compilers will catch type mismatches in function parameters. This can prevent unwanted type conversions.

clude traditional ones such as floating-point overflow, square root of a negative, divide-by-zero, and image related ones such as boundary violation, wraparound, and pixel overflow. The ability to define and handle exceptional conditions in the high-level language aids in the construction of interrupt handlers and other code used for real-time event processing. Moreover poor handling of exceptions can degrade performance. For example, floating-point overflow errors can propagate bad data through an algorithm and instigate time-consuming error recovery routines.

Of all the languages widely used in imaging applications, Ada has the most explicit exception handling facility, although ANSI-C provides some exception handling capability through the use of signals. Finally, exception handling can often be implemented in languages such as C, Pascal, and Modula-2 as a user-definable library when permitted by the compiler.

7.6 Modularity

Software engineers utilize different criteria for partitioning software systems into modules in a way that both clearly defines intermodule communication and that prevents unwanted intermodule interference. Application of these criteria in the design of real-time systems assures a reliable and maintainable design. In **Parnas partitioning** modules are defined so as to isolate hardware-dependent or volatile sections of code. The internal logic must be invisible to client modules but the interfaces are clearly defined.

For example, consider Fig 7.1. Here a graphics system is shown in hierarchical form. It consists of graphical objects (trees, houses, cars, and so on) that are composed from circles and boxes. Different objects can also reside in different display windows. Implementation of circles and boxes is based on the composition of line drawing calls. Thus, line drawing is the most basic and hardware-dependent function. Whether the hardware is based on pixel, vector, turtle, or other type of graphics does not matter – only the line drawing routine needs to be changed. Hence we have isolated the hardware dependencies to a single module. Certain languages have constructs that are designed to promote such techniques.

The concept of the `MODULE` is implicit in the language Modula-2. Each `MODULE` consists of a set of clearly defined input and output parameters, local and invisible variables, and a series of procedures. In Ada the notion of a package embodies the concept of Parnas information hiding exquisitely. The Ada package consists of a specification and declarations that include its

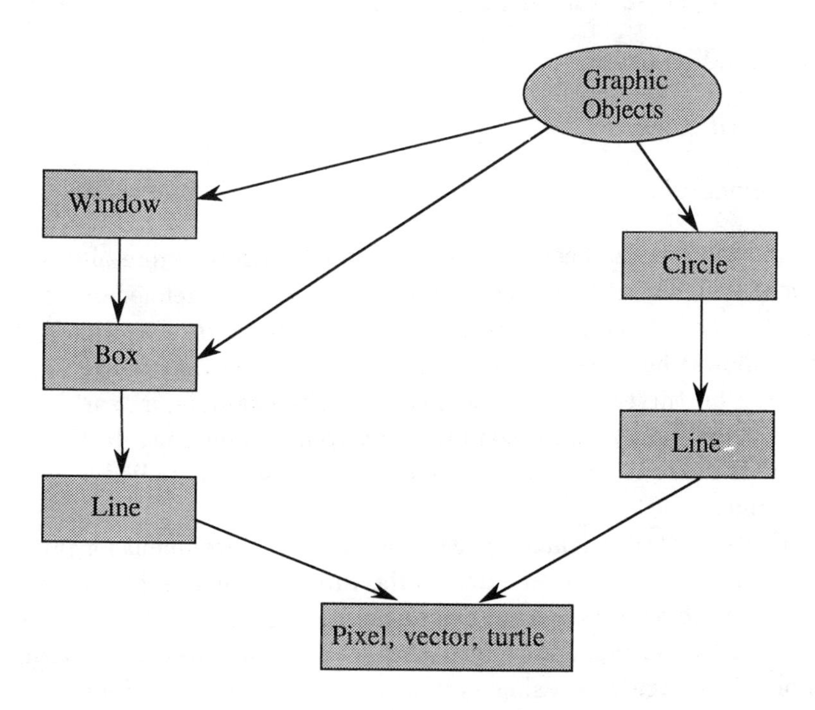

Figure 7.1: Parnas partitioning.

public or visible interface and its invisible or private parts. In addition the package body, which has further externally invisible components, contains the working code of the package. Packages are separately compilable entities, which further enhances their application as black boxes. In many versions of Pascal there is the notion of the separately compiled module called a `UNIT`. Similarly, in Fortran there is the notion of a `SUBROUTINE` and separate compilation of source files. These language features can be used to achieve modularity and design abstract data types. The C language also provides for separately compiled modules and other features that promote a rigorous top-down design approach that should lead to a good modular design.

While modular software is desirable, there is a price to pay in the overhead associated with procedure calls and parameter passing. This adverse effect should be considered when sizing modules.

7.7 Object-Oriented Languages

Formally, **object-oriented programming** languages must support

- data abstraction
- inheritance
- polymorphism.

Data abstraction has been previously defined. **Inheritance** allows the programmer to define new **objects** in terms of previously defined objects so that the new objects "inherit" characteristics or **attributes**. Function **polymorphism** allows the programmer to define operations that behave differently depending on the type of object involved. For example, a "sort" operation would act differently on integers than on roman numerals. But the request to "sort" is made in the same manner. How the sort is applied is determined at run time.

Object-oriented languages provide a natural environment for information hiding, which is similar to **encapsulation**. In encapsulation, a class of objects and the operations that can be performed on them (called **methods**) are enclosed or encapsulated into packages called **class definitions**. For example, in image processing systems, one may wish to define a class of type pixel, with characteristics (attributes) describing its position, color, brightness, and so on and operations that can be applied to a pixel such as add, activate, deactivate, and so on. One may then wish to define objects of type image as a collection of pixels with other attributes and so on. In some cases, expression of system functionality is easier to do in an object-oriented manner.

Object-oriented techniques are purported to increase programmer efficiency, reliability, and the potential for reuse. However, there is currently no hard evidence that this is so. Moreover, there is an execution time penalty to pay in real-time for object-oriented languages. The execution time penalty of object-oriented languages may in part be due to the relative immaturity of their compilers. But it seems clear that object-oriented requirements for late binding (resolution of memory locations at run-time rather than at compile time), necessitated by function polymorphism and inheritance, are irrevocable delay factors. Another problem results from the real-time collection of garbage generated by these types of languages. One possible way to reduce these penalties is not to define too many classes and only define classes that contain coarse detail and high-level functionality. But more research is needed in the application of object-oriented paradigms to real-time imaging systems.

7.8 Survey of Commonly Used Languages

In this section, we briefly discuss how certain widely used programming languages measure up for use in real-time imaging. It should be noted that there is no specialized language for image processing (meaning that it was designed specifically for this purpose) nor a language for image processing in real-time. An image processing language would include features that allow for description of and optimal storage management for image data types, provide exception handling for specific imaging exceptions (e.g., wraparound and clipping), and facilitate running on various parallel architectures. A real-time imaging language would allow for the aforementioned features and also permit the programmer to describe temporal behavior (e.g., when a process would run and how soon it could finish). And while there are some languages that have been designed specifically for real-time, most notably PEARL (a process control language used primarily in Germany), real-time Euclid (a variant of Pascal), and occam-2 (a multitasking language that can be used with transputers), these have not been widely used in imaging applications or have not entered the mainstream.

One final note. None of the languages discussed here generally supports parallel architectures. Hence our discussion is restricted to their use in von Neumann architectures.

Fortran

Fortran is the oldest high-level language still in use (ca. 1956). Because it has powerful formula expression, and most compilers generate fast code, it is still widely seen in legacy image processing software (software inherited from previous systems). Fortran has no explicit interrupt-handling capability and it is not modular in nature. Because dynamic allocation is not a standard feature of the language, images must be handled through static allocation of arrays. For these reasons and many others (two being that it is weakly typed and bit manipulation is not directly available), it is not recommended for use in real-time image processing.

BASIC

The BASIC language, which was long relegated to a teaching role in secondary schools, has made a strong resurgence through the appearance of visually oriented (and object-oriented) BASIC versions. BASIC, in its purest form (if there is one, as it is nonstandardized), is not at all suited for imaging or real-time. It was not developed for practical use and hence lacks most desirable features from a software engineering standpoint. It certainly is not desirable for interrupt handling and, in interpreted form, is much too slow. Some BASIC products can be used to call assembly language routines or code compiled by other languages. BASIC is then used to spawn tasks, handle the graphical user interface (GUI), and so forth, but not to do image processing.

Pascal

Pascal, which was also developed as a teaching language, is easy-to-use and learned in many universities. Thus it is often chosen for imaging applications. Pascal provides many desirable features for real-time imaging such as all forms of parameter passing, abstract data typing, and bit manipulation. However, Pascal compilers generate code that is notoriously slow and does not support interrupt handling. Hence, it is unsuitable for most real-time imaging applications.

C

The ANSI (American National Standards Institute) C programming language and its object-oriented descendent, C++, are probably the two most widely used programming languages for imaging. Perhaps this is so because C provides a "low-level" high-level programming interface, which allows for easy bit manipulations, "inline" directives that allow for embedded assembly code, and high-level control constructs like loops. In addition, abstract data typing, which is useful in imaging, is available through structures and unions. Furthermore, it is used in interrupt-driven systems because of its low-level nature, because interrupt handling is possible directly in the language without the need for assembly language, and because compilers for it tend to generate efficient code. The major drawback to the C language is that it lacks the special features needed for imaging and real-time that were previously discussed. Object-oriented C++ remedies this situation somewhat in that object classes for pixels, images, and so forth can be described along with methods relating to some of the image processing algorithms that we have discussed. However, C++ in most implementations is slower than C and does not support parallel architectures. Moreover, it has still to be shown that the heralded advantages of object-oriented languages (ease-of-use, reliability, lower design costs, etc.) are real. Nevertheless, C++ is the language of choice for many real-time imaging applications.

Ada

The Ada programming language (ca. 1982), which was mandated for all U.S. Department of Defense applications, has been designed specifically for embedded real-time systems. It is a highly modular, strongly typed language that supports exception and interrupt explicitly. It also provides abstract data typing that facilitates the creation of image types, and provides for explicit multitasking. Ada had many problems early in its life, especially in validating compilers that could generate efficient code. Moreover, some of the mechanisms that were provided for specifying timing behavior were found to be inadequate. A new standard for Ada, Ada 9X, is under development and is intended to rectify these shortcomings. However, many people find the Ada syntax restrictive and, if given a choice, avoid using the language.

Assembly Language

Assembly language, although lacking most of the high-level features desirable in real-time design, does have one advantage – more direct control of the hardware. Unfortunately, because of its unstructured nature and because it varies widely, coding in assembly language is usually difficult to learn, tedious, and error prone (try coding an FFT in assembly language; although it has been done, it is painful). The resulting code is also unportable. Nevertheless, many embedded systems use a small percentage of assembly language code for speed and efficiency. In short, assembly language programming should be limited to very tight timing situations or to control hardware features not supported by the compiler.

Understanding the mapping between high-level language input and the assembly language output from the compiler is essential. For example, in many C and Pascal compilers, the case statement is only efficient if more than three cases are to be compared – otherwise nested `if` statements generate better code. At other times, the code generated for a case statement can be quite complex. For example, if you have a C compiler, compare the code generated by the following two equivalent programs:

```
main()
{
  int pixel;
  void add(), delete();

  switch (pixel) {
    case 0:
    case 1:  add(pixel);
             break;
    case 2:
    case 3:
    case 4:  delete(pixel);
             break;
    case 5:
    case 6:
    case 7:  pixel = pixel-2;
             add(pixel);
             break;
    }
}

main()
{
  int pixel;
  void add(), delete();

  if (pixel == 0 || pixel == 1)
    add(pixel);
  else
    if(pixel == 2 || pixel == 3 || pixel == 4)
      delete(pixel);
```

```
  else
  {
    pixel = pixel - 2;
    add(pixel);
  };
}
```

For our compiler, the second form results in significantly better code (in terms of number of instructions). However, if the number of cases were to be increased, the case statement would be more efficient. Good compilers should provide optimization of the assembly language code output.

Chapter 8

Optimization Techniques

In general, to achieve optimal performance in image processing algorithms, designers attempt to match the hardware to the problem; for example, they use the parallel architectures discussed in Chapter 6. Failing this, a variety of techniques used in conjunction with high-level languages are employed to squeeze additional performance from the machine. These techniques include the use of assembly language patches and hand-tuning compiler output. Often, however, use of these practices leads to code that is unmaintainable and unreliable because it is poorly documented. More desirable, then, is to use coding tricks that involve direct interaction with the high-level language and that can be documented. These tricks improve real-time performance, but generally not at the expense of maintainability and reliability. Note that when optimizing average-case performance, worst-case performance is generally adversely affected.

In this chapter we discuss how to calculate one measure of real-time performance, CPU utilization. We then show how to use high-level languages in the best possible manner to achieve optimal performance.

8.1 CPU Utilization Estimation

CPU utilization estimates are measures that are meaningful primarily in cyclic real-time systems. In interrupt-driven systems, calculation of CPU utilization from measured data is easily computable only for periodic systems, to which we confine our discussion.

For periodic systems, CPU utilization is the sum of task execution times divided by the cycle times. In other fixed-rate, sporadic, or mixed systems, the maximum task execution period can be used in place of the cycle time. The CPU utilization T is given by

$$T = \sum_{i=1}^{n} \frac{A_i}{T_i}, \tag{8.1}$$

where n is the number of tasks, T_i is the cycle time (or minimum time between occurrences), and A_i is the actual execution time for task i. Recall that a CPU utilization of 100% or higher is considered a time-overloaded condition and will lead to missed deadlines.

For example, consider an imaging system where data are gathered from sensors every 5 milliseconds via an interrupt-driven routine (collecting the data takes 2.1 milliseconds). A second process, which is initiated every 30 milliseconds by an interrupt, processes the images and displays them on a CRT. This process requires 11 milliseconds to complete. Finally, a one-second rate process performs hardware diagnostics and takes 5 milliseconds to complete. The CPU utilization is then

$$2.1/5 + 11/30 + 5/1000 = 79.1\%.$$

8.2 Execution Time Estimation

As discussed above, CPU utilization is based on the execution time estimates for each procedure. But how do we measure or estimate execution time? The best method for measuring the execution time of any piece of code is to use a logic analyzer. One advantage of this is that hardware latencies and other delays not due simply to instruction execution times are taken into account. The drawback of the logic analyzer is that the system must be completely (or partially) coded and the target hardware available. Hence, the logic analyzer is usually only employed in the late stages of the coding phase, during the testing phase, and especially during system integration. The literature provided with the logic analyzer should indicate the specific technique.

When a logic analyzer is not available, the code execution time can be estimated by examining the compiler output and counting macroinstructions. This technique also requires that the code be written, an approximation of the final code exists, or similar systems are available for analysis. The approach simply involves tracing the worst-case path through the code, counting the macroinstructions along the way, and adding their execution times. These can be found in the manufacturer's specifications or through measurement with a logic analyzer.

Another accurate method of code execution timing uses the system clock, which is read before and after executing code. The time difference can then be measured to determine the actual time of execution. This technique, however, is only viable when the code to be timed is large relative to the code that reads the clock.

8.3 Basic Optimization Techniques

Identifying wasteful computation is crucial to reducing response times and CPU utilization. Many approaches used in compiler optimization can be used but methods have evolved that are specifically oriented toward real-time systems. We discuss those here.

Many of the basic techniques used in modern compiler design take advantage of well-known time-space trade-offs in computer systems. This means that if one attempts to reduce execution times, there is an increase in memory used and vice versa. Because imaging applications tend to be memory intensive, time-space techniques can be directly applied to imaging algorithms to save CPU cycles. These techniques include the following, which are discussed below:

- arithmetic identities
- constant folding
- reduction in strength
- common subexpression elimination
- loop invariant removal
- loop induction elimination
- loop unrolling
- loop jamming.

Arithmetic Identities

Common-sense application of arithmetic identities can reduce execution times significantly. For example, factorization and the avoidance of multiplications by 0 or 1 and addition by 0 can be advantageous. Sparse matrices such as those used for fast matrix transforms can particularly benefit from such an approach.

As an example, consider a pseudocode fragment for windowed convolution*:

```
for (r = -m to m) do
  for (s = -m to m) do
    h[i,j] = h[i,j] + g[r,s] * f[i+r,j+s]
```

One could replace the blind multiplication with a check if either multiplicand is 1 or 0 in the following manner:

```
for (r = -m to m) do
  for (s = -m to m) do
    if (g[r,s] <> 0 AND f[i+r,j+s] <> 0) then
      if (g[r,s] = 1) then
           h[i,j] = h[i,j] + f[i+r,j+s]
      else if (f[i+r,j+s] = 0) then
              h[i,j] = h[i,j] + g[r,s]
           else
              h[i,j] = h[i,j] + g[r,s] * f[i+r,j+s]
           endif
      endif
    endif
```

If the extra time needed to check the values of the multiplicands is less than the time needed to perform the multiplication, such a scheme will result in a performance improvement. This will be especially profound if the arithmetic operations are taking place in floating-point rather than as integers. Moreover, the above code is only for one pixel in the convolved image. The savings realized from this approach are compounded m^2 times if $i = j = m$. Finally, although we have reduced code execution time, we have increased the program size – a classic example of the time-space trade-off.

Arithmetic identities can benefit the algorithms for

*Recall that convolution of a signal with an impulse function yields the signal back, obviating the need for the convolution. A convolution algorithm might check for this case.

- Calculation of the Hadamard matrices, which requires only additions and subtractions.
- Calculation of the DCT
- Calculation of the DFT
- Application of the union operation in erosion.

One other variation of arithmetic identity should be considered as well – in most computers division takes longer than multiplication (a zero divisor must be checked for and division is implemented by negating the exponent and multiplying). Hence by replacing division with multiplication, modest savings can be achieved.

Constant Folding

An important rule in real-time code generation is to perform as many calculations as possible prior to run time. One example involves precomputing constants at compile time called **constant folding.** Constant folding generates additional constants, aptly illustrating the time-space trade-off. For example, many of the algorithms that we have discussed involve the constant 2π. (The imaginary number i cannot easily be represented with standard variable types; instead, two-dimensional arrays or structures are used where the first component is the real part and the second component is the imaginary part. We omit that consideration here.)

For example, we can create a variable called "pi," which will be initialized at compile time, and we use it in the expression

```
W = exp(-2 * pi / N)
```

If N is fixed, this code would be replaced with a new constant, which is evaluated at compile time as $e^{\frac{-2\pi}{N}}$, namely,

```
W = constant
```

Precomputation of constants is desirable anywhere combinations of constants are used frequently such as in

- The $\frac{1}{2}, \frac{1}{\sqrt{2}}, \frac{1}{\sqrt{8}}, \ldots$ constants used in the Hadamard matrices.

- Discrete Fourier coefficients $N^{\frac{-1}{2}}$, $e^{-2\pi}$.
- Discrete cosine coefficients $c_0(\frac{2}{N})^{\frac{1}{2}}$, $\frac{\pi}{N}$.

Although most compilers precompute constants invisibly and automatically, it is a good idea to do it explicitly, using high-level language code.

Reduction in Strength

Often an ALU provides several instructions that have the same effect. Increment, decrement, addition, and subtraction can all be used to control loop variables for example. Similarly, for integers, `SHIFT` and `MULT` instructions can be used in many of the same ways. Yet the execution time of a `SHIFT` instruction is faster than that of a `MULT` instruction. Persuading a compiler to replace arithmetic operations with equivalent, but faster operations is called **reduction in strength.**

For example, for any ordinal forms (such as integers, scaled numbers) multiplication or division by any power of 2 can be replaced with shift operations (which are generally faster). Similarly, division generally takes longer than multiplication; hence, we replace division with multiplication by the reciprocal. For example, multiply by 0.5 rather than divide by 2.0. These substitutions can be made by hand or by a smart compiler.

As a further implementation consider that multiplication of Hadamard matrices can be done without arithmetic multiplications; it requires only additions and subtractions (which in fact could be replaced with shifts).

Common Subexpression Elimination

Common subexpressions, especially those in loops, should be eliminated. And although most optimizing compilers will do this automatically, it is often safer, more efficient, and clearer to do this in the source code. As an example, consider the following pseudocode fragment:

```
z = pi * x + omega * y
q = omega * (pi + y)
```

Here we notice that `y * omega` represents a common subexpression and so we assign it to an intermediate variable. The resulting code is

```
t = y * omega
z = pi * x + t
q = pi + t
```

yielding a savings of one floating-point multiply. Generation of the intermediate variable **t** is essentially done at no cost since the values for **y** and **omega** must be loaded anyway, and **t** will most likely be retained in a register for the subsequent calculation, potentially providing a further savings by allowing the compiler to generate register direct instructions rather than direct mode.

Loop Invariant Removal

Optimizations involving loop constructs are very important since loops appear in many of the imaging applications that we have discussed. One loop optimization, called **loop invariant removal,** involves removing code that does not change inside a looping sequence. For example, consider the following code fragment from a DFT algorithm:

```
for (i = 1 to 100) do
  x[i] = x[i]+ 2 * omega * t
```

This code could be replaced by

```
z = 2 * omega * t
for (i = 1 to 100) do
  x[i] = x[i]+ z
```

thereby saving the needless recalculation of **2 * omega * t** inside the loop, which might entail two floating-point multiplies per iteration. Combining this technique with constant folding will save more time.

As another example of loop invariant removal, consider the code

```
for (i = 1 to 100) do
  x[i+1] = x[i+1] * pi
```

Here the loop invariant is not so obvious. It is, in fact, buried inside the array index. To remove the invariant, use the code

```
for (i = 2 to 101) do
  x[i] = x[i] * pi
```

By changing the looping rule, we have eliminated a needless addition inside the loop. It is worthwhile to examine all extant code for such opportunities.

Loop Induction Elimination

A variable i is called an **induction variable** if every time i changes its value, it is incremented or decremented by some constant. A common situation in image processing applications occurs when a linear function of the induction variable, j, is used to offset into some array and used for a loop termination test. We can derive a performance improvement by replacing the test for i with one for j. For example, we can replace the following pseudocode:

```
for (i = 1 to 1000) do
    image[i+1] = 0
```

with the code

```
for (j = 2 to 1001) do
    image[j] = 0
```

thus eliminating the extra addition within the loop, and possibly simplifying the addressing mode of the code generated.

Loop Unrolling

When calculating the performance of the windowed convolution in Chapter 3, we bypassed the effects of incrementing and testing a loop guard variable by coding directly in assembly language. If the code were written in a high-level language, however, this loop control code would be automatically generated. Loop control generally involves load register, test, jump, and increment instructions for each iteration. For example, consider pseudocode for the windowed convolution of Eq. 3.1:

```
for (r=-m to m) do
  for (s=-m to m) do
    h[i,j] = h[i,j] + g[r,s] * f[i+r,j+s]
```

The process of expanding a loop so that loop overhead is removed is called **loop unrolling** and it can save considerable execution time.

For example, in windowed convolution, if $m = 3$ and if we unroll the inner loop only, then the above pseudocode can be replaced by

```
for (r=-m to m) do
begin
  h[i,j] = h[i,j] + g[-3,-3] * f[i-3,j-3]
  h[i,j] = h[i,j] + g[-3,-2] * f[i-3,j-2]
  h[i,j] = h[i,j] + g[-3,-1] * f[i-3,j-1]
  h[i,j] = h[i,j] + g[-3,0] * f[i-3,j]
  h[i,j] = h[i,j] + g[-3,1] * f[i-3,j+1]
  h[i,j] = h[i,j] + g[-3,2] * f[i-3,j+2]
  h[i,j] = h[i,j] + g[-3,3] * f[i-3,j+3]
end
```

Removing the redundant reloading and storing `h[i,j]` (which a good compiler would do anyway) yields

```
for (r=-m to m) do
  h[i,j] = h[i,j] + g[-3,-3] * f[i-3,j-3] + g[-3,-2] * f[i-3,j-2]
                  + g[-3,-1] * f[i-3,j-1] + g[-3,0] * f[i-3,j]
                  + g[-3,1] * f[i-3,j+1]  + g[-3,2] * f[i-3,j+2]
                       + g[-3,3] * f[i-3,j+3]
```

This saves not only the overhead of managing the loop, but allows the compiler to generate instructions without being concerned with variable array indices. Furthermore, we can combine this approach with the arithmetic identities discussed previously, thus avoiding arithmetic operations involving multiplication by 0 or 1 and addition by 0 (the redundant load and store operations cannot be eliminated then). The combined effects of these approaches leads to code that is near the optimal for von Neumann architectures discussed in Chapter 3.

Loop unrolling is not always desirable if it leads to code that is difficult to test and maintain. However, in very-high-frequency computation cycles and for large images, noticeable savings may be realized. Finally, loop unrolling techniques can sometimes be used to transform linear algorithms into a form that is suitable for compilers that support parallel architectures. As this is implementation dependent, we cannot provide further details here.

Loop Jamming

Since looping is common in imaging applications, it is likely that two loops involving similar looping control will appear sequentially. If this is the case, one can combine the two loops within the control of one loop variable using the technique of **loop jamming.** As an example, consider a code fragment that might appear in a DCT algorithm:

```
for (i=1 to 100) do
  x[i] = y[i] * pi * 2

for (j=1 to 100) do
  z[j] = x[j+1] * pi * 2
```

This code can be replaced with

```
for (i=1 to 100) do
  begin
    x[i] = y[i] * pi * 2
    z[i] = x[i+1] * pi * 2
  end
```

Although the savings are usually modest relative to the calculations within the loop body, if the number of iterations is very large, the absolute time saved can be significant. Finally, loop jamming is not always possible in instances were the looping control appears similar. For example, the code previously shown for windowed convolution could not be jammed.

8.4 Combination Effects

Many of the aforementioned optimization techniques can be combined to provide additional execution time savings. Consider the following example, which illustrates this principle. The code fragment

```
for (j = 1 to 3) do
  begin
    a[j] = 0
    a[j] = a[j] + 2 * x
  end

for (k = 1 to 3) do
  b[k] = b[k] + a[k] + 2 * k * k
```

is improved by loop jamming, loop invariant removal, and removal of extraneous code (in this case the initialization of `a[j]`). The resultant code is

```
t = 2 * x
for (j = 1 to 3) do
  begin
    a[j] = t
    b[j] = b[j] + a[j] + 2 * j * j
  end
```

Next, loop unrolling yields

```
t = 2 * x
a[1] = t
b[1] = b[1] + a[1] + 2 * 1 * 1
a[2] = t
b[2] = b[2] + a[2] + 2 * 2 * 2
a[3] = t
b[3] = b[3] + a[3] + 2 * 3 * 3
```

Finally, after constant folding, the improved code is

```
t = 2 * x
a[1] = t
b[1] = b[1] + a[1] + 2
a[2] = t
b[2] = b[2] + a[2] + 8
a[3] = t
b[3] = b[3] + a[3] + 18
```

The original code involved 9 additions and 9 multiplications, numerous data movement instructions, and loop overheads. The improved code requires only 6 additions, 1 multiply, less data movement, and no loop overhead.

8.5 Scaled Numbers

In most computers, integer operations are faster than floating-point. We can exploit this by converting floating-point algorithms into scaled integer algorithms. In such a scheme, called **scaled numbers**, the least significant bit (LSB) of an integer variable is assigned a real number scale factor. Scaled numbers can be added and subtracted together and multiplied and divided by a constant (but not another scaled number). The results are converted to floating-point only at the last step – a process that can save considerable time.

For example, suppose an analog-to-digital converter is converting optical information from a camera into electrical impulses corresponding to gray values and storing them in a 16-bit unsigned integer. In this case, floating-point gray values in the range $0 \leq$ x < 1 indicate that the pixel is black (1) or white (0). If the least significant bit of the 16-bit integer has value 2^{-16}, then the most significant bit is $1 - 2^{-16}$. A common practice is to quickly convert the integer number into its floating-point equivalent by

```
xf = x * 0.0000153
```

and then proceed to use it in calculations directly with other converted numbers; for example,

```
diff = xf - zf
```

where `zf` is a similarly converted floating-point number. Instead, one can perform the calculation in integer form first and then convert to floating point:

```
diff = (x - z) * 0.0000153
```

For applications involving the manipulation, addition, and subtraction of large quantities of floating-point pixel data, scaled numbers can introduce significant savings. Note, however, that multiplication and division (by any number other than 0 or 1) cannot be performed on a scaled number as those operations change the scale factor. Finally, accuracy is generally sacrificed by excessive use of scaled numbers.

Another type of scaled number is based on the fact that adding 180° to any angle is analogous to taking its twos complement. This technique, called binary angular measurement, or BAM, works as follows. Consider the least significant bit of an n-bit word to be $2^{n-1} \cdot 180$ degrees with most significant bit (MSB) = 180 degrees. The range of any angle θ represented this way is

$$0 \leq \theta \leq 360 - 180 \cdot 2^{-(n-1)} \text{ degrees.}$$

A generic BAM word is shown in Fig. 8.1. For more accuracy, BAM can be extended to two more words. Binary angular measure is frequently used in navigation software and imaging systems using line manipulation algorithms such as ray tracing. In addition, it works well in conjunction with digitizing imaging devices.

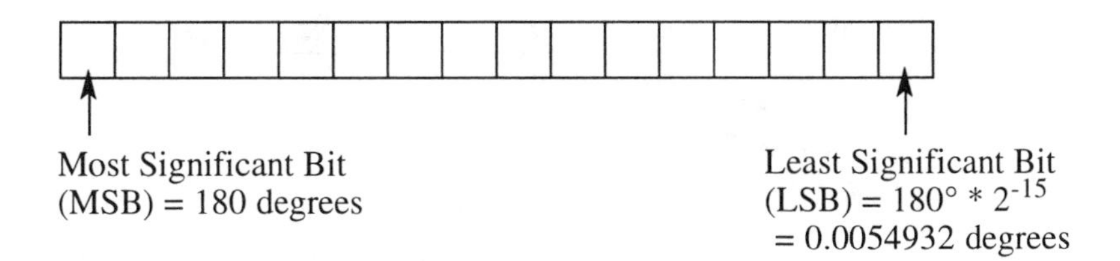

Ex:
0000 0000 1010 0110 = 166 * 180° * 2^{-15}= 0.9118°

Figure 8.1: A 16-bit binary angular measurement word.

8.6 Look-up Tables

Another variation of the scaled number concept uses a stored table of function values at fixed intervals. Such a table, called a **look-up table**, allows for the computation of continuous functions using mostly fixed-point arithmetic.

Let $f(x)$ be a continuous real function and let Δx be the interval size. We wish to store n values of f over the range $[x_0, x_0 + (n-1)\Delta x]$ in an array of scaled integers. Values for the derivative f' may also be stored in the table. The choice of Δx represents a trade-off between the size of the table and the desired resolution of the function. A generic look-up table is given in Table 8.1.

It is well known that the table can be used for the interpolation of $x < x' < x + \Delta x$ by the formula

$$f(x') = f(x) + (x' - x)\frac{f(x + \Delta x) - f(x)}{\Delta x}. \tag{8.2}$$

This calculation is done using integer instructions except for the final multiplication by the factor $\frac{(x'-x)}{\Delta x}$ and conversion to floating-point. As a bonus, the look-up table has faster execution time if x' happens to be one of the

x	$f(x)$
x_0	$f(x_0)$
$x_0 + \Delta x$	$f(x_0 + \Delta x)$
.	.
.	.
.	.
$x_0 + (n-1)\Delta x$	$f(x_0 + (n-1)\Delta x)$

Table 8.1: Generic look-up table.

table values. If $f'(x)$ is also stored in the table, then the look-up formula becomes

$$f(x') = f(x) + (x' - x)f'(x), \tag{8.3}$$

which improves the execution time of the interpolation somewhat, but increases memory requirements. The main advantage in using look-up tables is that if a table value is found and no interpolation is needed, then the algorithm is much faster than the corresponding series expansion. In addition, even if interpolation is necessary, the algorithm is interruptable and hence has a lower latency than a series expansion.

Look-up tables are widely used in the implementation of continuous functions like e^x, sine, cosine, tangent, their inverses, and so on. For example, consider the combined look-up table for sine and cosine using radian measure shown in Table 8.2. Because these trigonometric functions and exponentials are used frequently in conjunction with the DFT and DCT, look-up tables can provide considerable savings here.

8.7 Imprecise Computation

In cases where software routines are needed to provide mathematical support (in the absence of firmware support or coprocessors), complex algorithms are often employed to produce the desired calculation. For example, a Taylor series type expansion (perhaps using look-up tables for function derivatives) can be terminated early, at a loss of accuracy but with improved performance. Techniques involving early truncation of a series in order to meet deadlines are often called **imprecise computation**. Imprecise computation (sometimes called **approximate reasoning**) is often difficult to apply,

Angle (rads)	Cosine	Sine	Angle (rads)	Cosine	Sine
0.000	1.000	0.000	6.981	0.766	0.643
0.698	0.766	0.643	7.679	0.174	0.985
1.396	0.174	0.985	8.378	−0.500	0.866
2.094	−0.500	0.866	9.076	−0.940	0.342
2.793	−0.940	0.342	9.774	−0.940	−0.342
3.491	−0.940	−0.342	10.472	−0.500	−0.866
4.189	−0.500	−0.866	11.170	0.174	−0.985
4.887	0.174	−0.985	11.868	0.766	−0.643
5.585	0.766	−0.643	12.566	1.000	0.000
6.283	1.000	0.000			

Table 8.2: Look-up table for trigonometric functions.

however, because it is not always possible to determine what processing can be discarded, and at what cost.

A variation on imprecise computation occurs when applying an algorithm on a compressed version of the image rather than the image itself in order to save computation time. As an illustration of the approach, suppose a 256-gray-level document image has been captured and one desires to use a linear matched filter to identify characters within the image. Owing to the binary-like nature of most documents, one could proceed by first thresholding the document to a binary image and then applying a much faster sparse hit-or-miss matched filter.

8.8 Memory optimization

In modern computer architectures, memory constraints are not as troublesome as they once were. Nevertheless, in embedded applications or in legacy systems (those that are being reused), often the imaging engineer is faced with restrictions on the amount of memory available for program storage or for scratch-pad calculations, dynamic allocation, and so on. Since there is a fundamental trade-off between memory usage and CPU utilization (with rare exceptions), when one wishes to optimize for memory usage, it is necessary to trade performance to save memory. For example in the trigonometric function just discussed, using quadrant identities can reduce the need for a large look-up table, although the additional logic needed represents a small run-time penalty. In general, then, when optimizing for memory requirement reduction, do the opposite of the optimization techniques covered in this chapter.

Glossary

A

abstract data type a language construct where a user defines his or her own type (such as "pixel") along with the requisite operation that can be applied to it.

activity packet see **templates.**

address bus see **bus.**

approximate reasoning see **imprecise computa.**

assembly code programs written in **assembly language.**

assembly language a low-level programming language consisting primarily of symbolic representations of a processor's **macroinstruction** set.

associative memory a hardware scheme that employs a storage device that achieves multiple-data processing via a built-in search capacity. Also called **content-addressable memory.** See also **mask register** and **response store.**

asynchronous event an **event** that occurs at unpredictable points in the flow-of-control and is usually caused by external sources such as a clock signal.

attributes in **object-oriented programming** the characteristics of a **class definition** or **object.**

B

bus the wires that connect the **CPU** and main memory. The bus is used to exchange memory location information ("addresses") and data between the CPU and main memory in binary-encoded form. The width of the bus is determined by the number of bits or wires provided for the binary code. Usually the address and data wires are referred to as the **address bus,** and **data bus** respectively.

C

cache see **memory caching.**

cache hit ratio the percentage of time in which a requested instruction or data are actually in the **cache.**

call-by-address see **call-by-reference.**

call-by-reference a parameter-passing mechanism where the address of the parameter is passed by the calling routine to the called procedure so that it can be altered there. Also called **call-by-address.**

call-by-value a parameter-passing technique where the value of the actual parameter in the subroutine or function call is copied into the procedure's formal parameter.

canonical if E and F form a decomposition of **image** W, meaning $E \cup F = W$, then the pair (E, F) is said to be canonical.

center weighted median a type of **weighted median** filter that is found by repeating only the value at the window center.

central processing unit in a computer it provides for arithmetic and logical operations. Abbreviated **CPU**.

class definitions **object** declarations along with the **methods** associated with them.

CISC see **complex instruction set computer**.

compass gradients a set of eight **images** that when used in **windowed convolution** provides an **edge filter**.

complex instruction set computer a processor that is characterized by a large number of complex instructions involving long **microprograms**, numerous multilevel addressing modes, and sophisticated **CPUs**. Abbreviated **CISC**. Contrast with **RISC**.

compute-bound computations in which the number of operations is large in comparison to the number of I/O instructions.

constant folding an optimization technique that involves precomputing constants at compile time.

content-addressable memory see **associative memory**.

convolution see **windowed convolution**.

coprocessor a second, independent processor used to expand a **CPU's** **macroinstruction** set so that complicated operations need not be coded in high-level languages.

CPU see **central processing unit**.

CPU utilization a measure of the percentage of non-idle processing.

cycle stealing in **DMA** when bus contention between the **CPU** and other devices can occur, sometimes causing delays in the **fetch** stage of the fetch-

decode-execute cycle.

D

data bus see **bus.**

dataflow architectures an **MIMD** architecture where control flow is determined by the availability of data. See also **token, activity packet.**

DCT see **discrete cosine transform.**

decoding in a **CPU** determining which set of **microinstructions** corresponds to a given **macroinstruction.**

deterministic system a system where for each possible state, and each set of inputs, a unique set of outputs and the next state of the system can be determined. See **event determinism** and **temporal determinism.**

DFT see **discrete Fourier transform.**

digital image a function of two discrete variables.

dilation dual of **erosion.**

direct memory access an input/output scheme where access to the computer's memory is afforded to other devices in the system without **CPU** intervention. Abbreviated **DMA.** Contrast with **memory-mapped I/O** and **programmed I/O.**

direct mode a memory addressing scheme in which the operand is the data contained at the address specified in the address.

discrete cosine transform an image transform similar to the **discrete Fourier transform**, but which provides better separation for images with strong pixel-to-pixel correlation. Abbreviated **DCT**.

discrete Fourier transform an operation that separates an **image** into its frequency components. See also **Fast Fourier Transform**. Abbreviated **DFT**.

disjunctive normal form a representation of a Boolean expression that involves a logical sum of products (maximum of minima).

DMA see **direct memory access**.

double indirect mode a memory addressing scheme similar to **indirect mode** but with another level of indirection.

dynamic priority system a **preemptive priority system** where the task priorities can change during program execution. Contrast with **fixed priority system**.

E

edge filter an operation that takes in a gray-scale **image** and yields a binary image whose 1-valued pixels are meant to represent an edge within the original image.

embedded system a software system that does not have a generalized **operating system** interface and is used explicitly to control specialized hardware

enabled state in a **dataflow architecture** when all necessary **tokens** have arrived and the input lines are full. Also called the **ready state**.

encapsulation in **object-oriented programming**, a **class** of **objects** and the operations that can be performed on them (called **methods**) are enclosed or encapsulated into packages called **class definitions**.

erosion a basic morphological **image** operation used to construct binary filters. Its dual is **dilation.**

event any occurrence that results in a change in the sequential flow of program execution. See **asynchronous** and **synchronous event.**

event determinism means that the next state and outputs of a system are known for each set of inputs that trigger **events.**

execute in a **CPU** the process of acting upon a **microinstruction sequence.**

F

fast Fourier transform a fast version of the **discrete Fourier transform.** Abbreviated **FFT.**

FFT see **fast Fourier transform.**

fast Hadamard transform a way of improving compression using the **Hadamard matrix** via factorization.

fetching when the **CPU** retrieves **macroinstructions** from main memory.

firm real-time system a **real-time system** where some fixed small number of deadlines can be missed without total system failure.

fixed priority system a **preemptive priority system** where the task priorities cannot be changed once the system is implemented. Contrast with **dynamic priority system.**

fixed rate system a software system where **interrupts** occur only at fixed frequencies.

flat erosion see **moving minimum.**

flat dilation see **moving maximum.**

floating-point number a term describing the computer's representation of a real number.

flushed in **pipelining** when one of the instructions in the pipeline is a branch instruction and the prefetched instructions are removed because of obsolescence.

G

global variables those that are within the scope of all modules of the software system.

granulometry a sequence of **openings** using **structuring elements** that are of increasing size.

granulometric size distribution a distribution generated by counting the pixels in each succeeding filtered **image** using a **granulometry.**

gray-scale moving median filter a **moving median filter** for **images** that are not binary.

H

Hadamard matrix a special matrix used in a **lossy compression** scheme. See also **fast Hadamard transform.**

hard real-time system a **real-time system** where failure to meet even one deadline results in total system failure.

hit-or-miss transform in morphological **image** processing a **nonincreasing Boolean function** used for image restoration.

hypercube processor a processor configuration that is similar to the **linear array processor** except that each processor element communicates data along a number of other higher dimensional pathways.

I

indirect mode a memory addressing scheme in which the operand of the instruction is a memory address containing the effective address of the address (see Fig. 2.5).

immediate mode a memory addressing scheme that involves an integer operand that is usually contained in the next address after the instruction.

implied mode a memory addressing scheme that involves one or more registers that are implicitly defined in the operation determined by instruction.

imprecise computation techniques in which accuracy is sacrificed in order to meet deadlines. These techniques involving early truncation of a numerical series. Also called **approximate reasoning.**

increasing Boolean function see **positive Boolean function.**

induction variable a variable that every time its value changes, it is incremented or decremented by some constant.

inheritance in **object-oriented programming** a technique that allows the programmer to define new **classes** in terms of previously defined classes so that the new classes "inherit" characteristics or **attributes.**

interrupt a hardware signal that alters the sequential nature of the **fetch-decode-execute** cycle by transferring program control to special interrupt handler routines.

interrupt controller a device that provides additional **interrupt** handling capability to a **CPU.**

interrupt latency the inherent delay between when an **interrupt** occurs and when the **CPU** begins reacting to it.

interrupt register contains a bit map of all pending (latched) **interrupts.**

interrupt vector contains the identity of the highest-priority **interrupt** request.

J

Joint Photographic Experts Group a **lossy compression** technique that is an industry standard for **image** information storage and retrieval. Also called **JPEG.**

JPEG see **Joint Photographic Experts Group.**

L

linear array processor a processor architecture that has one PE for each column in the image, and enough memory directly connected to each PE to hold the entire column of image data for several distinct images. Also known as **vector processor.**

linear granulometries in a **granulometry**, the resulting **opening** sequences using vertical and horizontal line segments of increasing length.

locality-of-reference the notion that if you examine a list of recently executed program instructions, you will note that most of the instructions are localized to within a small number of instructions.

look-up table an optimization technique that allows for the computation of continuous functions using mostly fixed-point arithmetic.

loop invariant removal an optimization technique that involves removing code that does not change inside a looping sequence.

loop jamming an optimization technique that involves combining two loops within the control of one loop variable.

loop unrolling an optimization technique that involves expanding a loop so that loop overhead is removed.

lossless compression an **image** compression scheme where decompression yields the exact image that was compressed. Contrast with **lossy compression.**

lossy compression an **image** compression scheme where decompression yields an image that is not identical to the image that was compressed. Contrast with **lossless compression.**

M

macroinstructions the lowest level of user-programmable computer instructions.

mask register in **associative memory** a device that is used to block out certain bits in the comparand that are not to be checked in the retrieval process. In **interrupts** it contains a bit map either enabling or disabling specific **interrupts**.

matched filter a operation employing an **image** that gives back high values when a mask is located on top of the object of interest and lower values elsewhere.

memory caching a technique in which frequently used segments of main memory are stored in a faster bank of memory that is local to the **CPU** (called a **cache**).

memory-mapped I/O an input/output scheme where reading or writing involves executing a load or store instruction on a pseudomemory address mapped to the device. Contrast with **DMA** and **programmed I/O**.

mesh processor a processor configuration that is similar to the **linear array processor** except that each processor element also communicates data north and south.

methods in **object-oriented systems**, functions that can be performed on **objects**.

microcode see **microinstruction**.

microcontrollers a type of **von Neumann architecture** where there is no decoding of macroinstructions.

microinstruction a primitive instruction stored in **CPU** internal memory. Also called **microcode.**

microprogram a collection of **microinstructions** corresponding to a **macroinstruction.**

MIMD see **multiple instruction stream, multiple data stream.**

minimal representation for a **positive Boolean function** an equivalent representation where no product whose variable set does not contain the variable set of a distinct product can be deleted without changing the function.

minterm in **disjunctive normal form** a logical sum of products or conjunctions of Boolean variables is taken. These products are the minterms.

MISD see **multiple instruction, single-data.**

morphological gradient the nonlinear analog to a linear gradient.

morphological pattern spectrum the normalization of a **granulometric size distribution.**

moving average the average value of an **image** over the pixels in a window when it is centered at a certain pixel.

moving maximum an **image** operation found by translating the image to a pixel and then taking the maximum of all values in the translate. Contrast with **moving minimum.**

moving median a type of binary **image** filter used to suppress salt-and-pepper noise and preserve edges.

moving minimum an **image** operation found by translating the image to a pixel and then taking the minimum of all values in the translate. Contrast with **moving maximum.**

Motion Picture Engineers' Expert Group an **image** compression technique used for motion pictures. Also called **MPEG.**

MPEG see **Motion Picture Engineers' Expert Group.**

multimedia computing computing that involves computer systems with high-resolution graphics, CD-ROM drives, mice, high-performance sound cards, and **multitasking operating systems** that support these devices.

multiple instruction stream, single data stream a computer that can process two or more instructions concurrently on a single datum. Abbreviated **MISD.**

multiple instruction stream, multiple data stream a computer characterized by a large number of processing elements, each capable of executing numerous instructions. Abbreviated **MIMD.**

multiprocessing operating system an **operating system** where more than one processor is available to provide for simultaneity. Contrast with **multitasking operating system.**

multitasking operating system **an operating system** that provides sufficient functionality to allow multiple programs to run on a single processor so that the illusion of simultaneity is created. Contrast with **multiprocessing operating system.**

O

object in **object-oriented programming** an instance of a **class definition.**

object-oriented programming a programming style using languages that support **abstract data types, inheritance**, and function **polymorphism.**

opening a key morphological filter involving the union of translates.

operating system a collection of programs that control the resources of the computer.

P

parallel thinning a technique employing the **hit-or-miss transform** to restore noise-degraded images.

Parnas partitioning a software design technique where modules are defined so as to isolate hardware-dependent or volatile sections of code.

pipelining a rudimentary form of instruction concurrency is achieved through disjoint hardware to facilitate concurrent operations.

polymorphism in **object-oriented programming** a technique that allows the programmer to define operations that behave differently depending on the type of **object** involved.

positive Boolean function a Boolean function that can be represented as a logical sum of products in which no variables are complemented. Also called an **increasing Boolean function**

preempt a higher-priority task is said to preempt a lower-priority task if it **interrupts** the lower-priority task.

preemptive priority system an **operating system** that uses **preempt**ion schemes instead of **round-robin** or first-come-first-serve scheduling.

Prewitt gradient masks a set of two **images** that when used in **windowed convolution** provides an **edge filter**.

programmed I/O an input/output scheme where special **macroinstructions** are used to transfer data to and from the **CPU**. Contrast with **DMA**

and **memory-mapped I/O**.

Q

quadtree compression a **lossless compression** technique in which the **image** is divided recursively into four parts, stopping when each part is a constant value.

R

rate-monotonic system a **preemptive priority system** where task priorities are assigned so that the higher the execution frequency, the higher the priority.

reactive system a system in which functionality is driven by ongoing, **sporadic** interaction with its environment, such as in virtual reality.

ready state see **enabled state**.

real-time system a hardware and/or software system that must satisfy explicit bounded **response time** constraints to avoid failure. See **hard**, **firm**, and **soft** real-time system.

recursion a programming language feature whereby a procedure can call itself.

register direct mode a memory addressing scheme similar to **direct mode** except the operand is a **CPU** register and not an address.

register indirect mode a memory addressing scheme similar to **indirect mode** except the operand address is kept in a register rather than in another memory address.

reduced instruction set computer a special class of **SISD** architectures where a limited number of **macroinstruction** types and addressing modes simplify the decode and macroinstruction execution process. Abbreviated **RISC**. Contrast with **CISC**.

reduction in strength an optimization technique where the compiler is persuaded to replace arithmetic operations with equivalent, but faster operations.

response store in **associative memory** the tag memory used to mark memory cells.

response time the time between the presentation of a set of inputs and the appearance of all the associated outputs.

RISC see **reduced instruction set computer**.

Roberts gradient a **morphological gradient** calculated using a set of two **images** that when used in **windowed convolution** provides an **edge filter**.

round-robin system a **multitasking operating system** strategy in which each task is assigned a fixed time quantum in which to execute. Contrast with **preemptive priority system**.

run-length encoded a **lossless compression** technique where the **image** is stored as a sequence of triples specifying a pixel at which the image is black such that the pixel to the left is white for a specified the run length.

S

scaled number an optimization technique where the least significant bit (LSB) of an integer variable is assigned a real number scale factor.

secondary storage in computer devices such as hard disks, floppy disks, tapes, and so forth, that are not part of the physical address space of the CPU.

segment in **pipelining** a disjoint processing circuit. Also called a **stage.**

sequency coefficients the coefficients generated when multiplying an **image** by the **Hadamard matrix.**

single instruction stream, single data stream a type of computer where the **CPU** processes a single instruction at a time and a single datum at a time. Abbreviated **SISD.**

single instruction stream, multiple data stream a computer where each processing element is executing the same (and only) instruction, but on different data. Abbreviated **SIMD.**

SISD see **single instruction stream, single data stream.**

SIMD see **single instruction stream, multiple data stream.**

smoothing filter an operation applied to an **image** to suppress random additive pixel noise.

Sobel masks a set of two **images** that when used in **windowed convolution** provides an **edge filter.**

soft real-time system a real-time system where missing deadlines leads to performance degradation but not failure.

sporadic system a system with **interrupts** occurring sporadically.

stack filter **positive Boolean function** used as a filter in conjunction with **threshold sets.**

stage see **segment**.

starvation in **preemptive priority systems** when higher priority tasks prevent lower priority tasks from using needed resources.

status register in **interrupt** systems it contains the value of the lowest interrupt that will presently be honored.

strong-neighbor mask an **image** mask used in the **moving median filter**.

structuring element masking **image** used in **erosion**.

synchronous event an **event** that occurs at predictable times such as the execution of a conditional branch instruction or a hardware trap.

systolic processor a computer consisting of a set of interconnected cells, each capable of performing a simple operation and synchronized by an external clock or "heartbeat."

T

templates in a **dataflow architecture** a way of organizing data into **tokens**. Also called an **activity packet**.

temporal determinism in a deterministic system when the response time for each set of outputs is known.

threshold method a technique used in conjunction with the **Hadamard matrix** for selecting the compressed transform vector where those components with values above some threshold are kept. Contrast with the **zonal method**.

threshold set a way of partitioning an **image** into subimages that stack on top of each other, with each being a subset of the one below it. See also **stack filter**.

time-loading see **CPU utilization**.

time-overloaded system a software system that has a 100% or more **CPU utilization** factor.

token in **dataflow architectures** data items employed to represent the dynamics of a dataflow system.

transputer a fully self-sufficient, multiple instruction set **von Neumann processor** designed to be connected to other transputers.

V

vector processor see **linear array processor**.

von Neumann architecture a **CPU** employing a serial **fetch-decode-execute** process.

von Neumann bottleneck in a **von Neumann architecture** the fact that instructions or data can never be concurrently exchanged between main memory and the **CPU**.

W

waiting state in a **dataflow architecture** when the function cannot be executed since not all input lines contain a token.

watershed segmentation a **morphological gradient** edge detection scheme.

wavefront array processor similar to a **systolic processor** except that there is no external clock.

weighted median a type of **moving median** filter using integer weights $a_1, a_2, \ldots, a_m$. It is found by repeating a_i times the observation x_i, ordering the new set of $a_1 + a_2 + \ldots + a_m$ values, and then choosing the middle value as the output. See also **center weighted median.**

windowed convolution a mathematical operation where a mask of numerical weights defined over some window is translated across a digital **image** pixel by pixel, and at each pixel the arithmetic sum of products between the mask weights and the corresponding image pixels in the translated window is taken.

Z

zonal method a technique used in conjunction with the **Hadamard matrix** for selecting the compressed transform vector where one simply fixes the set of components to be kept. Contrast with the **threshold method.**

Bibliography

Chapter 1: What is Real-Time Processing?

Halang, W. A. and A. D. Stoyenko, *Constructing predictable real-time systems*, Kluwer Academic Publishers, Boston, 1991.

Laplante, P., *Real-Time Systems Design and Analysis: An Engineer's Handbook*, IEEE Press, Piscataway, NJ, 1992.

van Tilborg, A. and G.M. Koob, editors, *Foundations of Real-Time Computing*, vols. I, II, Kluwer Academic Publishers, Boston, 1991.

Chapter 2: Basic Hardware Architecture

Bartee, T. C., *Computer architecture and logic design.* New York: McGraw-Hill, 1991.

Dougherty, E. R., and C. R. Giardina, *Mathematical Methods for Artificial Intelligence and Autonomous Systems*, Prentice-Hall, Englewood Cliffs, 1988.

Hayes, J. P., *Computer architecture and organization.* 2nd ed. New York: McGraw-Hill, 1988, pp. 210-211.

Hill, F. J. and G. R. Peterson. *Digital systems: Hardware organization and design*, 3rd ed., New York, John Wiley & Sons, 1987.

Chapter 3: Linear Image Processing Algorithms

Dougherty, E. R., and C. R. Giardina, *Image Processing - Continuous to Discrete*, Prentice-Hall, Englewood Cliffs, 1987.

Gonzalez, R. C., and R. E. Woods, *Digital Image Processing*, Addison-Wesley, Boston, 1992.

Haralick, R. and L. Shapiro, *Machine Vision*, Addison-Wesley, Boston, 1991.

Jain, A. K., *Fundamentals of Digital Image Processing*, Prentice-Hall, Englewood Cliffs, 1989.

Lim, J. S., *Two-Dimensional Signal and Image Processing*, Prentice-Hall, Englewood Cliffs, 1990.

Pratt, W. K., *Digital Image Processing*, 2nd ed., John Wiley, New York, 1991.

Rosenfeld, A., and A. C. Kak, *Digital Picture Processing*, Academic Press, New York, 1992.

Chapter 4: Compression by Matrix Transforms

Brigham, E. O., *The Fast Fourier Transform and its Applications*, Prentice-Hall, Englewood Cliffs, 1988.

Clarke, R. J., *Transform Coding of Images*, Academic Press, New York, 1985.

Jones, P. W., and M. Rabbani, "Digital Image Compression," in *Digital Image Processing Methods*, ed. E. R. Dougherty, Marcel Dekker, New York, 1994.

Hsing, T. R., and A. G. Tescher, eds., *Selected papers on Visual Communication Technology and Applications*, SPIE Milestone Series, Vol. MS 13, SPIE Press, Bellingham, 1990.

Netravali, A. N., and B. G. Haskell, *Digital Pictures: Representation and Compression*, Plenum Press, New York, 1988.

Pennebaker, W. B., and J. L. Mitchell, *JPEG: Still Image Data Compression Standard*, Van Nostrand Reinhold, New York, 1992.

Rabbani, M., ed., *Selected Papers on Image Compression*, SPIE Milestone Series, Vol. MS 48, SPIE Press, Bellingham, 1992.

Rabbani, M., and P. W. Jones, *Digital Image Compression Techniques*, SPIE Press, Bellingmam, 1991.

Rao, K. R., and P. Yip, *Discrete Cosine Transform: Algorithms, Advantages, Applications*, Academic Press, San Diego, 1990.

Chapter 5: Nonlinear Image Processing Algorithms

Astola, J., and E. R. Dougherty, "Nonlinear Filters," in *Digital Image Processing Methods*, ed. E. R. Dougherty, Marcel Dekker, 1994.

Dougherty, E. R., *An Introduction to Morphological Image Processing*, SPIE Press, Bellingham, 1992.

Dougherty, E. R., ed., *Mathematical Morphology in Image Processing*, Marcel Dekker, New York, 1992.

Dougherty, E. R., and J. Astola, *An Introduction to Nonlinear Image Processing*, SPIE Press, Bellingham, 1994.

Giardina, C. R., and E. R. Dougherty, *Morphological Methods in Image and Signal Processing*, Prentice-Hall, Englewood Cliffs, 1988.

Pitas, I., and A. Venetsanopoulos, *Nonlinear Digital Filters*, Kluwer Academic Publishers,

Serra, J., *Image Analysis and Mathematical Morphology*, Academic Press, New York, 1982

Serra, J., ed., *Image Analysis and Mathematical Morphology*, Vol. 2, Academic Press, New York, 1988.

Chapter 6: Parallel Architectures

Duff, M. J. B., ed., *Computing Structures for Image Processing*, Academic Press, London, 1983.

Lawson, H. W., *Parallel Processing in Industrial Real-Time Applications*, Prentice-Hall, Englewood Cliffs, NJ, 1992.

Levialdi, S., ed., *Multicomputer Vision*, Academic Press, London, 1988.

Uhr, L., Preston, Jr., K., Levialdi, S., and M. J. B. Duff, eds., *Evaluation of Multicomputers for Image Processing*, Academic Press, New York, 1986.

Wilson, S. S., "Image-Processing Architectures," in *Digital Image Processing Methods*, ed. E. R. Dougherty, Marcel Dekker, New York, 1994.

Chapter 7: Programming Languages

Burns, A. and A. Wellings. *Real-time systems and their programming languages.* New York: Addison Wesley, 1990.

Horowitz, E., *Fundamentals of programming languages.* 2nd ed. Rockville, MD: Computer Science Press, 1984.

Laplante, P., *Real-Time Systems Design and Analysis: An Engineer's Handbook*, IEEE Press, Piscataway, NJ, 1992.

Chapter 8: Optimization Techniques

Aho, A., Sethi R., and J. Ullman. *Compilers: Principles, techniques and tools.* New York: Addison Wesley, 1986.

Jain, R., *The Art of Computer Systems Performance Analysis*, John Wiley, New York, 1991.

Laplante, P., *Real-Time Systems Design and Analysis: An Engineer's Handbook*, IEEE Press, Piscataway, NJ, 1992.

Index

Edward R. Dougherty is a professor at the Center for Imaging Science of the Rochester Institute of Technology and also serves as an industrial consultant. He holds an M.S. in computer science from Stevens Institute of Technology and a Ph.D. in mathematics from Rutgers University. He has written numerous professional papers on image processing and has authored/co-authored eight books, including the present *SPIE* tutorial text and two other *SPIE* tutorial texts, *An Introduction to Nonlinear Image Processing* (with Jaakko Astola) and *An Introduction to Morphological Image Processing*. He has also edited two books in the Marcel Dekker Optical Engineering Series: *Mathematical Morphology in Image Processing* and *Digital Image Processing Methods.* He is currently the editor of the SPIE/IS&T *Journal of Electronic Imaging*, serves as associate editor for the journals of *Mathematical Imaging and Vision* and *Real-Time Imaging,* serves as chair of the SPIE Working Group on Electronic Imaging, serves as chair for two SPIE conferences, *Image Algebra and Morphological Image Processing* and *Nonlinear Image Processing,* and regularly teaches conference short courses in morphological, nonlinear, and real-time imaging.

Phillip A. Laplante (Senior Member, IEEE) received a B.S. degree, an M.Eng., and a Ph.D. in Computer Science, Electrical Engineering, and Computer Science, respectively, from the Stevens Institute of Technology. He is currently an Associate Professor and Chair in the Department of Mathematics and Computer Science at Fairleigh Dickinson University (Madison, New Jersey) and a visiting researcher at the Real-Time Systems Laboratory at New Jersey Institute of Technology. Dr. Laplante's interests are in image processing, real-time systems, and real-time image processing, including virtual reality. He has authored over three dozen technical papers and six books. He is the founding co-editor-in-chief of the new journal, *Real-Time Imaging.* Dr. Laplante is a licensed Professional Engineer in the State of New Jersey, a member of ACM, SPIE, and the IEEE Press Editorial Board.